Invasion Biology

Hypotheses and Evidence

———————————————

For our families

CABI Invasives Series

Invasive species are plants, animals or microorganisms not native to an ecosystem, whose introduction has threatened biodiversity, food security, health or economic development. Many ecosystems are affected by invasive species and they pose one of the biggest threats to biodiversity worldwide. Globalization through increased trade, transport, travel and tourism will inevitably increase the intentional or accidental introduction of organisms to new environments, and it is widely predicted that climate change will further increase the threat posed by invasive species. To help control and mitigate the effects of invasive species, scientists need access to information that not only provides an overview of and background to the field, but also keeps them up to date with the latest research findings.

This series addresses all topics relating to invasive species, including biosecurity surveillance, mapping and modelling, economics of invasive species and species interactions in plant invasions. Aimed at researchers, upper-level students and policy makers, titles in the series provide international coverage of topics related to invasive species, including both a synthesis of facts and discussions of future research perspectives and possible solutions.

Titles Available

1. *Invasive Alien Plants: An Ecological Appraisal for the Indian Subcontinent*
   Edited by J.R. Bhatt, J.S. Singh, S.P. Singh, R.S. Tripathi and R.K. Kohli

2. *Invasive Plant Ecology and Management: Linking Processes to Practice*
   Edited by T.A. Monaco and R.L. Sheley

3. *Potential Invasive Pests of Agricultural Crops*
   Edited by J.E. Peña

4. *Invasive Species and Global Climate Change*
   Edited by L.H. Ziska and J.S. Dukes

5. *Bioenergy and Biological Invasions: Ecological, Agronomic and Policy Perspectives on Minimizing Risk*
   Edited by L.D. Quinn, D.P. Matlaga and J.N. Barney

6. *Biosecurity Surveillance: Quantitative Approaches*
   Edited by F. Jarrad, S. Low-Choy and K. Mengersen

7. *Pest Risk Modelling and Mapping for Invasive Alien Species*
   Edited by Robert C. Venette

8. *Invasive Alien Plants: Impacts on Development and Options for Management*
   Edited by Carol A. Ellison, K.V. Sankaran and Sean T. Murphy

9. *Invasion Biology: Hypotheses and Evidence*
   Edited by Jonathan M. Jeschke and Tina Heger

Invasion Biology

Hypotheses and Evidence

_____

Edited by

Jonathan M. Jeschke[1,2,3] and Tina Heger[4,5,3]

[1]*Leibniz-Institute of Freshwater Ecology and Inland Fisheries (IGB), Berlin, Germany*

[2]*Freie Universität Berlin, Institute of Biology, Berlin, Germany*

[3]*Berlin-Brandenburg Institute of Advanced Biodiversity Research (BBIB), Berlin, Germany*

[4]*University of Potsdam, Biodiversity Research/Systematic Botany, Potsdam, Germany*

[5]*Technical University of Munich, Restoration Ecology, Freising, Germany*

CABI is a trading name of CAB International

CABI
Nosworthy Way
Wallingford
Oxfordshire OX10 8DE
UK

CABI
745 Atlantic Avenue
8th Floor
Boston, MA 02111
USA

Tel: +44 (0)1491 832111
Fax: +44 (0)1491 833508
E-mail: info@cabi.org
Website: www.cabi.org

T: +1 (617)682-9015
E-mail: cabi-nao@cabi.org

The designations employed and the presentation of the material in this publication do not imply the expression of any opinion whatsoever on the part of CABI concerning the legal status of any country, territory, city of area or of its authorities, or concerning the delimitation of its frontiers or boundaries. Lines on maps represent approximate border lines for which there may not yet be full agreement.

A catalogue record for this book is available from the British Library, London, UK.

Library of Congress Cataloging-in-Publication Data

Names: Jeschke, Jonathan M., editor. | Heger, Tina, editor.
Title: Invasion biology : hypotheses and evidence / edited by Jonathan M. Jeschke & Tina Heger.
Description: Boston, MA : CABI, [2018] | Series: CABI invasives series; 9 | Includes bibliographical references and index.
Identifiers: LCCN 2017040053 (print) | LCCN 2017049378 (ebook) | ISBN 9781780647654 (ePDF) | ISBN 9781780647661 (ePub) | ISBN 9781780647647 (hbk: alk. paper)
Subjects: LCSH: Biological invasions--Research.
Classification: LCC QH353 (ebook) | LCC QH353 .I5763 2018 (print) | DDC 577.7/18--dc23
LC record available at https://lccn.loc.gov/2017040053

ISBN-13: 9781800621619 (pbk)/9781780647647 (hbk)

Commissioning editor: David Hemming
Editorial assistant: Emma McCann
Production editor: Ali Brown

Typeset by AMA DataSet Ltd, Preston, UK.
Printed and bound in the UK by Severn, Gloucester.

Previously published in Hardback 2018.

Contents

Contributors

―――――――――――――

Amador-Vargas, Sabrina, Smithsonian Tropical Research Institute, Balboa, Panamá. E-mail: samadorv@gmail.com

Blackburn, Tim M., Department of Genetics, Evolution & Environment, Centre for Biodiversity & Environment Research, UCL, Gower Street, London WC1E 6BT, UK; Institute of Zoology, Zoological Society of London, Regent's Park, London NW1 4RY, UK; School of Biological Sciences and the Environment Institute, The University of Adelaide, North Terrace SA 5005, Australia. E-mail: t.blackburn@ucl.ac.uk

Braga, Raul Rennó, Programa de Pós-Graduação em Ecologia e Conservação, Universidade Federal do Paraná, Curitiba, Brazil; Laboratório de Ecologia e Conservação, Depto de Engenharia Ambiental, Setor de Tecnologia, Universidade Federal do Paraná, 81531-970, Curitiba, Paraná, Brazil. E-mail: raulbraga@onda.com.br

Cassey, Phillip, School of Biological Sciences and the Environment Institute, The University of Adelaide, North Terrace SA 5005, Australia. E-mail: phill.cassey@adelaide.edu.au

Debille, Simon, Freie Universität Berlin, Institute of Biology, Königin-Luise-Str. 1-3, 14195 Berlin, Germany; KU Leuven - University of Leuven, Department of Biology, Leuven, Belgium. E-mail: simon.debille@gmail.com

Enders, Martin, Freie Universität Berlin, Institute of Biology, Königin-Luise-Str. 1-3, 14195 Berlin, Germany; Leibniz-Institute of Freshwater Ecology and Inland Fisheries (IGB), Berlin, Germany; Berlin-Brandenburg Institute of Advanced Biodiversity Research (BBIB), Berlin, Germany. E-mail: enders.martin@gmx.net

Erhard, Felix, Technical University of Munich, Restoration Ecology, 85350 Freising, Germany; University of Natural Resources and Life Sciences (BOKU), Vienna, Austria. E-mail: felix.erhard@googlemail.com

Farji-Brener, Alejandro G., Lab. Ecotono, INIBIOMA, Universidad Nacional del Comahue and CONICET, Bariloche, Argentina. E-mail: alefarji@yahoo.com

Fox, Gordon A., Department of Integrative Biology (SCA 110), University of South Florida, 4202 E. Fowler Ave, Tampa, FL 33620-2000, USA. E-mail: gfox@usf.edu

García-Díaz, Pablo, School of Biological Sciences and the Environment Institute, The University of Adelaide, North Terrace SA 5005, Australia; Landcare Research New Zealand, PO Box 69040, Lincoln 7640, New Zealand. E-mail: garcia-diazp@landcareresearch.co.nz

Gómez Aparicio, Lorena, Department of Biogeochemistry, Plant and Microbial Ecology, Institute of Natural Resources and Agrobiology of Seville (IRNAS), CSIC, Sevilla, Spain. E-mail: lorenag@irnase.csic.es

Griesemer, James, University of California, Davis, Philosophy, One Shields Avenue, Davis, CA 95616-8673, USA; University of California, Davis, Center for Population Biology, Davis, CA, USA. E-mail: jrgriesemer@ucdavis.edu

Heger, Tina, University of Potsdam, Biodiversity Research/Systematic Botany, Maulbeerallee 1, 14469 Potsdam, Germany; Technical University of Munich, Restoration Ecology, Freising, Germany; Berlin-Brandenburg Institute of Advanced Biodiversity Research (BBIB), Berlin, Germany. E-mail: tina-heger@web.de

Jeschke, Jonathan M., Freie Universität Berlin, Institute of Biology, Königin-Luise-Str. 1-3, 14195 Berlin, Germany; Leibniz-Institute of Freshwater Ecology and Inland Fisheries (IGB), Berlin, Germany; Berlin-Brandenburg Institute of Advanced Biodiversity Research (BBIB), Berlin, Germany. E-mail: jonathan.jeschke@gmx.net

Lockwood, Julie L., Department of Ecology, Evolution and Natural Resources, Rutgers University, 14 College Farm Road, New Brunswick, NJ 08901-8525, USA. E-mail: julie.lockwood@rutgers.edu

Lortie, Christopher J., York University, Department of Biology, 4700 Keele St., Toronto, ON M3J 1P3, Canada. E-mail: lortie@yorku.ca

Müller, Caroline, Department of Chemical Ecology, Bielefeld University, Universitätsstr. 25, 33615 Bielefeld, Germany. E-mail: caroline.mueller@uni-bielefeld.de

Nordheimer, Regina, Freie Universität Berlin, Institute of Biology, Königin-Luise-Str. 1-3, 14195 Berlin, Germany. E-mail: regina-nordheimer@hotmail.de

Pyšek, Petr, Institute of Botany, Department of Invasion Ecology, The Czech Academy of Sciences, 252 43 Průhonice, Czech Republic; Charles University, Faculty of Science, Department of Ecology, Prague, Czech Republic. E-mail: pysek@ibot.cas.cz

Scheiner, Samuel M., Div. Environmental Biology, U.S. National Science Foundation, 2415 Eisenhower Ave., Alexandria, VA 22314, USA. E-mail: sscheine@nsf.gov

Starzer, Julian, Technical University of Munich, Restoration Ecology, 85350 Freising, Germany. E-mail: julian-starzer@t-online.de

Torchyk, Olena, Technical University of Munich, Restoration Ecology, 85350 Freising, Germany. E-mail: olenatorchyk@hotmail.de

Vitule, Jean Ricardo Simões, Laboratório de Ecologia e Conservação, Depto de Engenharia Ambiental, Setor de Tecnologia, Universidade Federal do Paraná, 81531-970, Curitiba, Paraná, Brazil; Programa de Pós-Graduação em Ecologia e Conservação, Universidade Federal do Paraná, Curitiba, Brazil. E-mail: biovitule@gmail.com

Foreword

_____

The story behind this book starts about a decade ago, in late 2007, when the two of us met for the first time. It was at a workshop that TH organized at the Technical University of Munich, Germany, and at which JMJ participated. We have enjoyed our joint projects ever since. Two of these have led to the formulation of the hierarchy-of-hypotheses (HoH) approach, namely our work on Jeschke *et al.* (2012) and discussions during the workshop 'Tackling the emerging crisis of invasion biology: How can ecological theory, experiments, and field studies be combined to achieve major progress?' (March 2010 in Benediktbeuern, Germany; workshop of the specialist group 'Theory in Ecology' of the Ecological Society of Germany, Austria and Switzerland, GfÖ), organized by TH, Sylvia Haider, Anna Pahl and JMJ (see Heger *et al.*, 2013). In Jeschke *et al.* (2012), we have used this approach to classify and analyse empirical tests of six major invasion hypotheses, finding only mixed empirical support. We further developed the HoH approach in Heger and Jeschke (2014) with a focus on the enemy release hypothesis.

In addition to dividing a given major hypothesis into more specific sub-hypotheses, it has become clear that a tool for connecting existing invasion hypotheses would be very useful. Such a tool could serve as a map for invasion biology, where the major hypotheses are landmarks such as cities in an ordinary map of a country. A network of hypotheses could thus potentially work as such a tool (Jeschke, 2014).

We therefore decided to take the next logical step with this book: to combine the HoH approach with hypothesis networks for invasion biology. A book allows for outlining the approach, suggestions and challenges much more than a series of papers published in different journals over different years. We used this book to further develop the HoH approach by inviting critical comments (Part I), apply it to 12 major invasion hypotheses (Part II) and explore how it can be expanded to a hierarchically structured hypothesis network (Chapter 7 and Part III). But the book does not come alone. It is important to also check out the companion website that features, for instance, data from the about 1100 studies analysed in Part II of the book: www.hi-knowledge.org. Importantly and as outlined in Chapter 17, this website will be further updated and developed, and we envision it will become a powerful web portal for invasion biology and other research disciplines in the future. The latter is possible because the HoH approach and hypothesis networks are not at all restricted to invasion biology but are applicable across all research disciplines.

This book contains 18 chapters written by 23 authors from 13 countries. Although being an edited volume and thus benefiting from the combined expertise of its multiple authors with diverse backgrounds, the book has also elements of a monograph because we were involved in most chapters. This involvement hopefully made sure that the book follows a clear structure and consistent line of thought, which is a frequent challenge for edited volumes. We hope that this mix of a monograph and edited volume is a good read, and that you won't regret you got a copy of it.

The chapters in this book were peer-reviewed. JMJ handled the reviewing process; only if he was an author and TH was not, TH handled the manuscripts. We reviewed all chapters that we did not write ourselves (editorial review). For every chapter, there was at least one additional reviewer who remained anonymous unless he/she opted to reveal their name. Chapters authored by both of us were reviewed by two referees. In this way, each chapter was reviewed by 2–3 experts. Reviewers were either authors of other book chapters or other experts on the topic. Chapters 2 to 6 in the book followed a different review process: Chapter 2 was openly reviewed in Chapters 3–5, and we respond to the points raised in these open reviews in Chapter 6.

We very much thank Maud Bernard-Verdier, Raul Braga, Martin Enders, Franz Essl, James Griesemer, Reuben Keller, Christopher Lortie, Caroline Müller, Martin Nuñez, Conrad Schittko, Margaret Stanley, David Strayer, Mark Torchin, Meike Wittmann and Florencia Yannelli for taking the time to read and carefully review the chapters. We also highly appreciate the open reviews presented in Chapters 3–5 and the comments provided by Jane Catford on hypothesis characteristics included in Chapter 7. All comments we received were helpful and contributed to the final contents of this book – many, many thanks! We also thank CABI for the opportunity to publish this book and David Hemming for initiating it. Finally, JMJ acknowledges financial support from the Deutsche Forschungsgemeinschaft (DFG; JE 288/9-1).

The book does not have to be read from the beginning to the end. For example, if you are mainly interested in our suggestions on how to improve research synthesis, in the hierarchy-of-hypotheses (HoH) approach and hypothesis networks, you can focus on Parts I and III of the book and only read Chapter 7 in Part II. If, however, you are mainly interested in one or two particular invasion hypotheses, you can pick the relevant chapter(s) in Part II of the book. If you would like to get an overview of existing invasion hypotheses and a synthesis of related empirical studies, you may go to Chapters 7 and 17.

We wish you an enjoyable read and would highly appreciate any constructive feedback you might have.

With our best wishes,
Jonathan M. Jeschke and Tina Heger
jonathan.jeschke@gmx.net; tina-heger@web.de
Berlin and Potsdam, August 2017

References

Heger, T. and Jeschke, J.M. (2014) The enemy release hypothesis as a hierarchy of hypotheses. *Oikos* 123, 741–750.

Heger, T., Pahl, A.T., Botta-Dukat, Z., Gherardi, F., Hoppe, C., Hoste, I., Jax, K., Lindström, L., Boets, P., Haider, S. *et al.* (2013) Conceptual frameworks and methods for advancing invasion ecology. *Ambio* 42, 527–540.

Jeschke, J.M. (2014) General hypotheses in invasion ecology. *Diversity and Distributions* 20, 1229–1234.

Jeschke, J.M., Gómez Aparicio, L., Haider, S., Heger, T., Lortie, C.J., Pyšek, P. and Strayer, D.L. (2012) Support for major hypotheses in invasion biology is uneven and declining. *NeoBiota* 14, 1–20.

Part I

Introduction to Invasion Biology and the Hierarchy-of-hypotheses Approach

_____

1

Invasion Biology: Searching for Predictions and Prevention, and Avoiding Lost Causes

Phillip Cassey,[1,*] Pablo García-Díaz,[1,2] Julie L. Lockwood[3] and Tim M. Blackburn[1,4,5]

[1]*School of Biological Sciences and the Environment Institute, The University of Adelaide, Australia;* [2]*Landcare Research New Zealand, Lincoln, New Zealand;* [3]*Department of Ecology, Evolution and Natural Resources, Rutgers University, New Brunswick, USA;* [4]*Department of Genetics, Evolution & Environment, Centre for Biodiversity & Environment Research, UCL, UK;* [5]*Institute of Zoology, Zoological Society of London, Regent's Park, London, UK*

Abstract

The introduction and establishment of alien species is one of the many profound influences of ongoing anthropogenic global environmental change. Invasion biology has emerged as the interdisciplinary study of the patterns, processes and consequences of the redistribution of biodiversity across all environments and spatio-temporal scales. The modern discipline hinges on the knowledge that biological invasions cannot be defined and studied solely by their final outcome of establishing alien species but rather as a sequential series of stages, or barriers, that all alien species transit: the 'invasion pathway'. Some of the most important influences for a species transiting these sequential stages (i.e. transport, introduction, establishment and spread) are event-level effects, which vary independently of species and location, such as the number of individuals released in any given location (propagule pressure). The number of studies of biological invasions has increased exponentially over the past two decades, and we now have a significant body of research on different aspects of the invasion process. In particular, the hierarchical nature of the invasion pathway has lent itself strongly to modern statistical methods in hierarchical modelling. Now, the science behind invasive species management must continue to develop innovative ways of using this accumulated knowledge for delivering actionable management procedures.

Introduction

Human populations have had a profound impact on all natural environments and the biological diversity they contain and support (Magurran, 2016). Alongside massive population growth, the continued rapid urbanization and globalization of human technology, transport and trade are all increasing this impact. In response, new scientific disciplines have emerged to evaluate the patterns, processes and consequences of human-induced global environmental change (Costanza *et al.*, 2007). One such discipline is invasion biology: the study of species (populations and individuals) that have

* Corresponding author. E-mail: phill.cassey@adelaide.edu.au

been redistributed outside their native geographic ranges as a result of human-mediated translocation (hereafter termed 'alien species'). Alien species can be transported to their new recipient locations intentionally through deliberate introduction or accidentally through unintentional 'stowaway' and/or escape (Hulme, 2009).

Throughout human history the translocation of species has greatly benefited human welfare and livelihood. Today, there are many tens-of-thousands of alien species that are used in foodstuffs (farmed, cultivated and harvested), as commercial materials (timber, packaging, clothing, derivatives and pharmaceuticals), and as ornamentals (garden plants, pets and commensal species). In rare cases (and not without controversy) alien species have also been promoted to assist species conservation (Schlaepfer *et al.*, 2011) and for responding to biodiversity loss through anthropogenic climate change (i.e. via assisted translocation; McLachlan *et al.*, 2007). Nevertheless, the ledger of the effects of alien species also has a long (arguably longer) debit column. Invasive alien species (IAS) act as a massive drain on global resources (Early *et al.*, 2016). Recent assessments suggest that they cost the economy of the UK at least £1.7 billion per annum (Williams *et al.*, 2010) and that of the EU at least €12.5 billion per annum (pa) and probably substantially more. Estimates of economic costs for other regions are similarly high (Pimentel *et al.*, 2000). These economic debits stem from the loss of productivity in agriculture, aquaculture and forestry; mitigation and control costs associated with new building construction, energy utilities and transportation infrastructure; and the costs of surveillance, quarantine and eradication efforts designed to prevent known invasive alien species' establishment in new regions (Kettunen *et al.*, 2008). Alien species are also one of the major drivers of biodiversity loss worldwide. IAS have contributed to the extinction of more plants and animals since 1500 AD than any process other than habitat loss (Bellard *et al.*, 2016). The widespread establishment of the same sets of alien species around the world erodes evolutionarily distinct communities (floras and faunas) by the process of biotic homogenization (Lockwood and McKinney, 2001). The ongoing establishment of new alien species, from almost all major taxa (Seebens *et al.*, 2017), suggests that society is facing increasing economic and ecological costs from IAS in the coming decades (Essl *et al.*, 2011).

The great variety of alien species, the widespread locations in which they have been released and become invasive pests, and the range of negative impacts arising from these pests has spawned a major research effort designed to understand (i.e. quantify and clarify) the invasion process, and provide the robust evidence-based activities needed to control and mitigate their impacts (Lockwood *et al.*, 2013).

The Invasion Pathway

The genesis of invasion biology as a scientific discipline is debatable but is often argued to have started with the publication of Charles Elton's book *The Ecology of Invasions of Animals and Plants* (Elton, 1958). The field nevertheless languished for a quarter of a century or so after this publication, surviving mostly within the applied work of entomologists, rangeland ecologists, weed scientists and wildlife biologists (Baker, 1974; Davis, 2006). This situation began to change when the Scientific Committee on Problems in the Environment (SCOPE) published a series of books and articles on the subject in the early 1980s (Drake *et al.*, 1989; Davis, 2006). The SCOPE participants identified three questions for invasion biology to answer:

1. What factors determine whether a species becomes an invader or not?
2. What site properties determine whether an ecological system will resist or be prone to invasions?
3. How should management systems be developed to best advantage given the knowledge gained from studying questions 1 and 2?

Although the SCOPE programme led to many national initiatives and influential edited volumes (Simberloff, 2011), the first few decades of research into biological invasions following SCOPE were characterized by a failure to make significant progress in addressing these questions in either an explanatory or a predictive manner (Williamson, 1993, 1999).

This scientific drought ended with the widespread recognition that biological invasions should not be defined and studied solely by their final outcome of producing alien species, but rather as a sequential series of stages, or barriers, that alien species transit (Williamson, 1993). These stages – commonly categorized as transport, introduction, establishment and spread (Fig. 1.1) – constitute the invasion pathway (also known as the invasion process). Only a (native) species that successfully overcomes all of the biogeographical, social, demographic, environmental and dispersal barriers, throughout the invasion stages, will become an invasive alien species (Blackburn et al., 2011). These stages differ in the nature of the barriers imposed, and therefore the

mechanisms required to overcome them. Notably, each stage generates a different set of hypotheses for how a species might transition through it (Kolar and Lodge, 2001). Partitioning the process of invasion into these stages required the testing of ideas about which species succeed at each stage in the process and which fail.

Transport

Changes in the mode and frequency by which species are transported have greatly affected temporal patterns in the sources and subsequent distribution of alien species. Early human transportation by foot, horse and sailing ship have been replaced by high-speed – and high-volume – rail, truck, ship and plane transport (Essl et al., 2015). The networks these transportation vectors travel have themselves expanded exponentially over the past century, making the species native to newly 'opened' regions subject to becoming transported out as aliens and making the regions themselves subject to

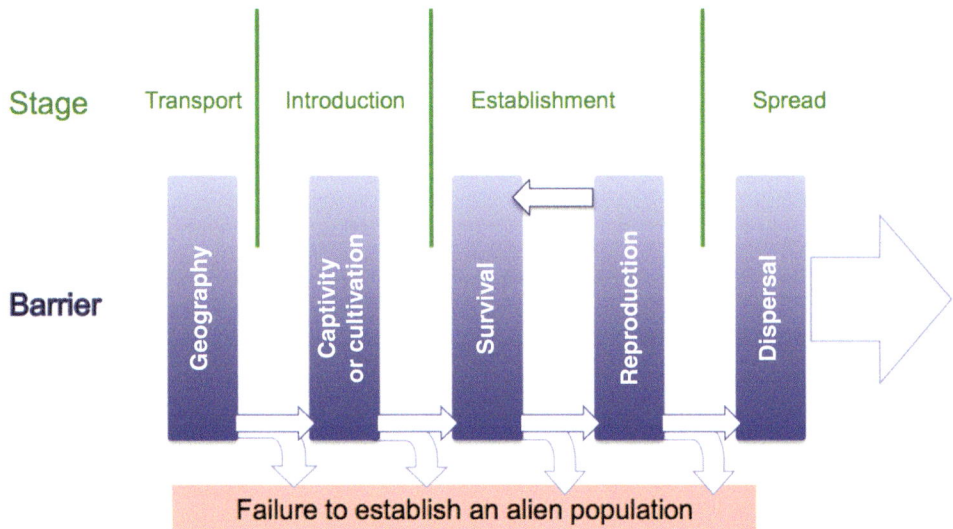

Fig. 1.1. A depiction of the sequential series of stages or barriers that define the invasion pathway. This framework identifies that the invasion process can be divided into a series of stages and that in each stage there are barriers that need to be overcome for a species or population to pass on to the next stage. Modified from Blackburn et al. (2011).

becoming invaded (Seebens *et al.*, 2015). The technological advancement of transportation and the globalization of trade has thus resulted in a massive expansion of the range of taxa moved by transport vectors and the speed with which these species are moved from their origin to their recipient locations (Ruiz and Carlton, 2003; Essl *et al.*, 2015). Commodities such as fresh produce (e.g. fruit, vegetables and flowers), pets (e.g. turtles, scorpions and groupers), and ornamentals (e.g. aquatic plants, marine algae and live coral) can now be sent from nearly anywhere to nearly anywhere else in the world in less than 24 hours. Along with these commodities come their transmissible pests and diseases, and the smaller species that stowaway either on the commodity itself or in its packaging. It is clear when analysing different periods of invasion biology that different taxa and regions predominate, and that these temporal patterns strongly reflect changes and advances in global trade (Dyer *et al.*, 2017). Recognition of the dynamic and expansive mechanisms of alien species transport forced invasion biologists to develop research programmes centred on the locations and types of species most likely to be entrained, and the causes and consequences of transport and vector dynamics (e.g. Essl *et al.*, 2015; Turbelin *et al.*, 2017).

Introduction

Not all individuals (or species) that are transported outside of their native range are introduced (released) into the recipient (alien) range. In some cases they do not survive transportation and thus never leave the vector (e.g. ship ballast, cargo hold), and in other cases they are contained in captivity once off-loaded into the alien range (e.g. pets, ornamental plants). Although these contained species still pose a potential invasion risk (Hulme, 2011; Cassey and Hogg, 2015), they cannot establish until they are introduced into the recipient environment. The factors that determine which species

make up the pool that are initially entrained in a transportation vector are poorly understood. Similarly, there is a paucity of research into the characteristics that facilitate either their survival within the vector until release or eventual release from captivity.

The introduction stage represents a critical target for the successful management of alien species, yet we lack an appropriate level of understanding of its dynamics. Transport and introduction are sometimes combined in the same stage for theoretical or practical analytical reasons (Leung *et al.*, 2012), which might have contributed to our limited understanding of the introduction stage. It is crucial, however, that we disaggregate these two stages, because transiting from transported to introduced breaks the major containment barrier for alien species (e.g. captivity), and potentially allows establishment (Hulme, 2011; Cassey and Hogg, 2015). Despite the sparseness of research on the likelihood of introduction, the shared conclusion across transport vectors and taxa is that increasing trade volumes of a commodity increases the likelihood of an introduction (García-Díaz *et al.*, 2017). Interestingly, this finding links well with the propagule pressure hypothesis explaining species' success in the next stage, establishment (see also Chapter 16, this volume). Species' traits effectively play a role in facilitating whether a species transits from transport to introduction (Wonham *et al.*, 2001; Su *et al.*, 2016; Vall-llosera and Cassey, 2017). Unfortunately, there remains a substantial knowledge gap regarding this aspect. The introduction of alien species is a multi-faceted process compounded by a multitude of factors including human behaviours (e.g. reasons why people release pets or dispose of unwanted ornamental plants into natural habitats) (Cohen *et al.*, 2007; Dehnen-Schmutz *et al.*, 2007). Accordingly, deciphering the function of different factors in determining the likelihood of introductions is a challenging task requiring a multidisciplinary research agenda incorporating elements from the ecological, economical and social sciences.

Establishment

The majority of research effort on the invasion pathway, to date, has been focused on the factors affecting the successful establishment of alien populations. Whether or not an introduction results in a population becoming self-sustaining depends on the intersection of three broad categories of driver: species-, location- and event-level characteristics (Duncan *et al.*, 2003). The first two of these categories echo the first two questions of the SCOPE programme above. Species-level factors include hypotheses on how life history, evolutionary history and genetic diversity influence the probability that a nascent alien population will become self-sustaining. The location-level characteristics include the myriad hypotheses on how (and if) disturbance leaves locations more likely to be invaded, and the role of inter-specific interactions in determining the fate of a recently introduced alien population (competition, predation, mutualism), among others. The event-level category reflects barriers that vary independently of species and location, such as the number of individuals released in any given location (e.g. propagule pressure; Lockwood *et al.*, 2005; Simberloff, 2009b). The event-level category was not identified and thus not addressed within the SCOPE programme (Blackburn *et al.*, 2015). Hypotheses for establishment success now tend to partition along the lines of these three categories or the intersection between them. For example, underlying the variability in species, events and environments is the common relationship between a species ability to survive in a new location (abiotic and biotic tolerances) and the demographic capability of the population to increase (R_0; e.g. Cassey *et al.*, 2014).

Spread

Not all established alien species spread widely across their available range but some manage to occupy vast expanses, even occasionally spreading over entire continents (Lockwood *et al.*, 2013). Key hypotheses invoke the role of landscape-level habitat patterns, the presence and strength of inter-specific interactions, and species traits that promote dispersal and phenotypic evolution or plasticity. An interesting phenomenon at this stage relates to those species that have apparent lag periods between establishment and spread, or so-called sleeper-species (Crooks, 2005; Aikio *et al.*, 2010; Aagaard and Lockwood, 2014). It is currently unknown whether these species-by-environment combinations are different from alien populations that initially expand and spread. Similarly perplexing is why some species, once established and spreading (seemingly beyond a small-population extirpation threshold), suddenly decline and 'disappear' (i.e. 'boom and bust'; Simberloff and Gibbons, 2004; Cooling *et al.*, 2011; Aagaard and Lockwood, 2016). In both these cases, local adaptation may be involved. In the first case of lag periods, the alien species might require a period of time to adapt to the new environment and, in the case of boom and bust, some feature of the new environment (e.g. a predator or pathogen) might become adapted to the alien (Blackburn and Ewen, 2017), with negative consequences. Hypotheses concerning evolutionary and disease dynamics are likely to be fruitful avenues to explore and improve understanding of alien species spread (or lack thereof). They will need to explain, however, why aliens are displacing native species that have apparently had orders of magnitude more generations to perfect their responses to local environmental conditions (i.e. the so-called 'paradox of invasion'; Sax and Brown, 2000; Fridley *et al.*, 2007).

Impacts

The aspect of alien species invasions that first garnered attention to the field was their realized ecological and economic impact.

Most recent conceptualizations of the invasion pathway, however, conspicuously avoid invoking impact as an invasion stage (e.g. Blackburn *et al.*, 2011; Jeschke *et al.*, 2013). There are three reasons for this decision. First, the presence and degree of impact is inevitably shaped by human perception, especially when it is defined using socio-economic metrics (Lockwood *et al.*, 2013). This prohibits the designation of 'impact' as a purely biological phenomenon, and thus makes difficult the task of formulating testable hypotheses around what ecological or evolutionary conditions favour high versus low impact (Thomsen *et al.*, 2011a). Second, even when impact is defined as a purely ecological outcome of an alien invasion, it has proven difficult to define it in a consistent and thus testable manner. Recognizing existing differences in human perspectives, when addressing socio-economic impacts, is vital for framing the discussion on alien species in the broader society. Third, alien species can have impacts at any stage of the invasion pathway (Jeschke *et al.*, 2014).

In a wide-ranging review of impact hypotheses, Ricciardi *et al.* (2013) defined impact as any measurable change to the properties of an ecosystem by an alien species. These properties can be measured at any biological scale of organization from genes to ecosystem functions, and the change imposed by the alien species can be either positive or negative as measured on an ecological (and not socio-economic) scale. Thus, for example, a zebra mussel population established as an alien within a river basin may increase water clarity, or it may over-grow native mussels at the same location and decrease their populations, sometimes to the point of local extinction (Strayer, 2009). Whether either of these measurable ecological changes are 'good' or 'bad' depends exclusively on the values one brings to this judgement (Lockwood *et al.*, 2013). By defining impact in a purely ecological context, Ricciardi *et al.* (2013) allowed for hypotheses on ecological impact to be generated and rigorously tested.

Even with this definition in place, there remain several hurdles to understanding when and where alien species impose ecological impacts. The first was recognized by Parker *et al.* (1999) when they disaggregated mechanisms of impact into three classes: those that are (i) regulated by the abundance of the alien species; (ii) the geographical range size of the alien; or (iii) the per-capita effect of the alien on co-occurring species or the ecosystem. There have been several synthetic reviews of alien species impacts in recent years; however, most have not attempted to identify which of these factors most influences impact levels across species or ecosystems, in part because not all classes of mechanism are equally easy to measure (e.g. Vilà *et al.*, 2011; Maggi *et al.*, 2015). A second hurdle comes from recognizing that invasive alien species may have some populations that have high impact and others that have low or even no impact (Cameron *et al.*, 2016). In addition, these levels of impact are not always static through time, raising the possibility that a single alien population can manifest impacts in a single ecosystem that range from acute to minimal depending on how long it has been resident there (Strayer *et al.*, 2006).

At least 13 hypotheses have been developed to explain why some alien species have strong impacts on co-occurring species, often dominating entire assemblages, while others remain relatively benign. The hypotheses proffered thus far reflect those commonly evoked to explain variation in establishment success and community ecology (Ricciardi *et al.*, 2013). Thus, impacts are supposed to vary according to traits of the alien species itself, its functional distinctiveness relative to co-occurring native species, the level of disturbance and environmental heterogeneity of the invaded ecosystem, and the strength and number of inter-specific interactions (Ricciardi *et al.*, 2013). These hypotheses are not mutually exclusive, and in most instances where an alien population imposes large impacts on ecosystems more than one mechanism appears to be working simultaneously (Thomsen *et al.*, 2011b). More recently, there has been interest in documenting whether spread and impact are always correlated with one another, and if so, what is the shape of this relationship (Yokomizo *et al.*, 2009).

If an alien population shows strong and negative impacts, defined either ecologically or socio-economically (or both), then invasion biologists have sought ways to inform the process of eradicating or controlling them. The science behind invasive species management must develop ways of using the accumulated knowledge from testing hypotheses of invasion from above into actionable management procedures (Dibble *et al.*, 2013). As one might expect, the degree to which managers have been successful at realizing measurable success in eradication or control of invasive species is mixed (Dibble *et al.*, 2013).

Eradication of established alien species is difficult and costly, and for widespread species can be simply impossible (Myers *et al.*, 2000). No widespread continental population of an established vertebrate invasive alien species has ever been successfully eradicated, although human-mediated extinctions of many native species suggest that it is possible. Nevertheless, eradications of small or geographically circumscribed populations are a successful management tool and, combined with sufficient planning and resources (including public and political support), can be effectively implemented (Howald *et al.*, 2007). When eradication is infeasible, managers move toward a programme of control whereby the invasive alien population is reduced in size or extent to protect valued assets (Simberloff, 2009a). These assets may be ecologically valuable, such as reducing the impacts of invasive plants or vertebrates within protected areas, or they may be economically or socially valuable, such as when reducing the populations of alien insect pests on crops. Future research in the implementation and conduct of new and sometimes controversial technologies for the eradication and control of alien populations (e.g. gene-editing approaches; Piaggio *et al.*, 2017) will need to be developed, in concert with methods to circumvent potential sociological conflicts that potentially render invasions 'wicked problems' for management (Ricciardi *et al.*, 2017).

The Trail Beyond

Invasion biology has come of age, in leaps and bounds, in recent years. Just over two decades ago, leading practitioners were positing that the field might simply be inherently unpredictable (Williamson, 1999). Since then the widespread adoption of the pathway model, the recognition that different drivers apply at different stages, and the incorporation of propagule pressure as a null model for establishment success, has brought us to the point where invasion biology as a predictive science can be envisioned (if not yet realized). Studies of biological invasions have increased exponentially over the last two decades (Simberloff, 2011; Lockwood *et al.*, 2013). The ongoing acceleration in the establishment of alien species has led to (and been identified by) an explosion of data on the occurrence and spread of alien populations (Dyer *et al.*, 2016, 2017; Seebens *et al.*, 2017), and robust research in biosecurity planning and preparedness (Bacon *et al.*, 2012; Banks *et al.*, 2015; Cope *et al.*, 2016). We have seen a flowering of hypotheses proposed to explain the distributions – and success or failure – of alien species, structured to align with the conceptual advances in partitioning of the invasion process; 35 of these hypotheses are featured in Chapter 7, and 12 hypotheses are outlined in detail in Chapters 8–16 of this volume. The challenges facing invasion biology have evolved and changed.

Biological invasions are inherently complex. They intersect the effects of human history, societal imperatives, environmental vagaries and evolutionary histories. It is likely that the stages on the invasion pathway (and the way we manage them) each have complex explanations and are determined by multiple drivers. As such, we do not expect single simple explanations for any given stage. Nevertheless, we also do not expect every hypothesis in invasion biology to be correct. Some will be viable, whereas others are zombies: dead but not buried. The key now is to distinguish between the two so we can be shorn of the dead. Much of invasion biology is based on observational data from unplanned and poorly designed 'experiments in nature' (as defined by Diamond and Case, 1986). In the past, this restricted our science to the analysis of small comparative datasets and comparisons of single (anecdotal) studies. This is no longer the case. We are now confronted with an increasingly large (in space and time) set of catalogues, compendiums and globally vast databases of research on biological invaders. Given that we have a significant body of studies on different aspects of the invasion process, one good way to move forward is to interrogate this information attempting to identify consensus, or key aspects of variation, in our existing knowledge.

Developments in statistical approaches led by medical science means that we have the research structures in place to examine bodies of evidence in favour of hypotheses in a formative, logical and quantitative manner. Systematic meta-analyses provide one particular tool, which can be usefully adopted for assessing and contrasting the strength of statistical effect-sizes across a range of studies, including the overwhelming variety of alien species, invaded locations and ecological outcomes (e.g. Sorte *et al.*, 2010; Vilà *et al.*, 2011). In addition, the hierarchical nature of the invasion pathway lends itself strongly to modern statistical methods in hierarchical modelling. Such approaches can be adopted to test for heterogeneity in the strength of effect sizes, assisting in determining which moderator variables underlie observed variability in invasion stage outcome. Quantitative meta-analyses can also be useful for resolving whether certain results (effects) are confounded by the potential pseudo-replication of dominant authors within the field, or of multiple studies being conducted on 'well-worn' datasets (e.g. Hulme *et al.*, 2013). A key area for development may involve refining the questions on which meta-analytical methods are brought to bear. So, for example, we may find that studies that we collectively consider as addressing the same idea actually concern several (albeit related) hypotheses (the hierarchy-of-hypotheses approach outlined

in Chapters 2 and 6, this volume). Alternately, complex definitions for common hypotheses can lead to a wide variety of measurements for the same underlying trait, such as the case for IAS impact (Kumschick *et al.*, 2015). The growing number of studies using proxy measurements for interpreting the relationship between propagule pressure and the different successful transitions in the invasion pathway are a good example of this as well (Wonham *et al.*, 2013). Clear refinement of the hypotheses and analysis of the underlying heterogeneity in these studies is an obvious benefit of robust and reproducible quantitative syntheses, which will define the way we address invasion biology in future years.

References

Aagaard, K. and Lockwood, J. (2014) Exotic birds show lags in population growth. *Diversity and Distributions* 20, 547–554.

Aagaard, K. and Lockwood, J.L. (2016) Severe and rapid population declines in exotic birds. *Biological Invasions* 18, 1667–1678.

Aikio, S., Duncan, R.P. and Hulme, P.E. (2010) Lag-phases in alien plant invasions: separating the facts from the artefacts. *Oikos* 119, 370–378.

Bacon, S.J., Bacher, S. and Aebi, A. (2012) Gaps in border controls are related to quarantine alien insect invasions in Europe. *PloS ONE* 7, e47689.

Baker, H. G. (1974) The evolution of weeds. *Annual Review of Ecology and Systematics* 5, 1–24.

Banks, N.C., Paini, D.R., Bayliss, K.L. and Hodda, M. (2015) The role of global trade and transport network topology in the human-mediated dispersal of alien species. *Ecology Letters* 18, 188–199.

Bellard, C., Cassey, P. and Blackburn, T.M. (2016) Alien species as a driver of recent extinctions. *Biology Letters* 12, 20150623.

Blackburn, T. M. and Ewen, J. G. (2017) Parasites as drivers and passengers of human-mediated biological invasions. *EcoHealth* 14 (Suppl. 1), 61–73.

Blackburn, T.M., Pyšek, P., Bacher, S., Carlton, J.T., Duncan, R.P., Jarošík, V., Wilson, J.R. and Richardson, D.M. (2011) A proposed unified framework for biological invasions. *Trends in Ecology & Evolution* 26, 333–339.

Blackburn, T.M., Dyer, E., Su, S. and Cassey, P. (2015) Long after the event, or four things we (should) know about bird invasions. *Journal of Ornithology* 156, 15–25.

Cameron, E. K., Vilà, M. and Cabeza, M. (2016) Global meta-analysis of the impacts of terrestrial invertebrate invaders on species, communities and ecosystems. *Global Ecology and Biogeography* 25, 596–606.

Cassey, P. and Hogg, C. J. (2015) Escaping captivity: the biological invasion risk from vertebrate species in zoos. *Biological Conservation* 181, 18–26.

Cassey, P., Prowse, T.A. and Blackburn, T.M. (2014) A population model for predicting the successful establishment of introduced bird species. *Oecologia* 175, 417–428.

Cohen, J., Mirotchnick, N. and Leung, B. (2007) Thousands introduced annually: the aquarium pathway for non-indigenous plants to the St Lawrence Seaway. *Frontiers in Ecology and the Environment* 5, 528–532.

Cooling, M., Hartley, S., Sim, D.A. and Lester, P.J. (2011) The widespread collapse of an invasive species: Argentine ants (*Linepithema humile*) in New Zealand. *Biology Letters*, rsbl20111014.

Cope, R.C., Ross, J.V., Wittmann, T.A., Prowse, T.A. and Cassey, P. (2016) Integrative analysis of the physical transport network into Australia. *PloS ONE* 11, e0148831.

Costanza, R., Graumlich, L., Steffen, W., Crumley, C., Dearing, J., Hibbard, K., Leemans, R., Redman, C. and Schimel, D. (2007) Sustainability or collapse: what can we learn from integrating the history of humans and the rest of nature? *Ambio* 36, 522–527.

Crooks, J.A. (2005) Lag times and exotic species: the ecology and management of biological invasions in slow-motion. *Ecoscience* 12, 316–329.

Davis, M.A. (2006) Invasion biology 1958–2005: the pursuit of science and conservation. In: Cadotte, M.W., McMahon, S.M. and Fukami, T. (eds) *Conceptual Ecology and Invasion Biology: Reciprocal Approaches to Nature*. Springer, Dordrecht, the Netherlands, pp. 35–64.

Dehnen-Schmutz, K., Touza, J., Perrings, C. and Williamson, M. (2007) The horticultural trade and ornamental plant invasions in Britain. *Conservation Biology* 21, 224–231.

Diamond, J. and Case, T. (1986) *Community Ecology*. Harper & Row, New York.

Dibble, K.L., Pooler, P.S. and Meyerson, L.A. (2013) Impacts of plant invasions can be reversed through restoration: a regional meta-analysis of faunal communities. *Biological Invasions* 15, 1725–1737.

Drake, J.A., Mooney, H.A., Di Castri, F., Groves, R.H., Kruger, F.J., Rejmánek, M. and Williamson, M. (1989) *Biological Invasions: a Global Perspective*. Wiley, New York.

Duncan, R.P., Blackburn, T.M. and Sol, D. (2003) The ecology of bird introductions. *Annual Review of Ecology, Evolution, and Systematics* 34, 71–98.

Dyer, E.E., Franks, V., Cassey, P., Collen, B., Cope, R.C., Jones, K.E., Şekercioğlu, Ç.H. and Blackburn, T.M. (2016) A global analysis of the determinants of alien geographical range size in birds. *Global Ecology and Biogeography* 25, 1346–1355.

Dyer, E.E., Cassey, P., Redding, D.W., Collen, B., Franks, V., Gaston, K.J., Jones, K.E., Kark, S., Orme, C.D.L. and Blackburn, T.M. (2017) The global distribution and drivers of alien bird species richness. *PLOS Biology* 15, e2000942.

Early, R., Bradley, B.A., Dukes, J.S., Lawler, J.J., Olden, J.D., Blumenthal, D. M., Gonzalez, P., Grosholz, E.D., Ibañez, I., Miller, L.P. and Sorte, C.J. (2016) Global threats from invasive alien species in the twenty-first century and national response capacities. *Nature Communications* 7, 12485.

Elton, C. S. (1958) *The Ecology of Invasions by Animals and Plants*. Methuen, London.

Essl, F., Dullinger, S., Rabitsch, W., Hulme, P.E., Hülber, K., Jarošík, V., Kleinbauer, I., Krausmann, F., Kühn, I., Nentwig, W. *et al.* (2011) Socioeconomic legacy yields an invasion debt. *Proceedings of the National Academy of Sciences USA* 108, 203–207.

Essl, F., Bacher, S., Blackburn, T.M., Booy, O., Brundu, G., Brunel, S., Cardoso, A.-C., Eschen, R., Gallardo, B., Galil, B. *et al.* (2015) Crossing frontiers in tackling pathways of biological invasions. *BioScience* 65, 769–782.

Fridley, J., Stachowicz, J., Naeem, S., Sax, D., Seabloom, E., Smith, M., Stohlgren, T., Tilman, D. and Holle, B.V. (2007) The invasion paradox: reconciling pattern and process in species invasions. *Ecology* 88, 3–17.

García-Díaz, P., Ross, J.V., Woolnough, A.P. and Cassey, P. (2017) Managing the risk of wildlife disease introduction: pathway-level biosecurity for preventing the introduction of alien ranaviruses. *Journal of Applied Ecology* 54, 234–241.

Howald, G., Donlan, C., Galván, J.P., Russell, J. C., Parkes, J., Samaniego, A., Wang, Y., Veitch, D., Genovesi, P. and Pascal, M. (2007) Invasive rodent eradication on islands. *Conservation Biology* 21, 1258–1268.

Hulme, P.E. (2009) Trade, transport and trouble: managing invasive species pathways in an era of globalization. *Journal of Applied Ecology* 46, 10–18.

Hulme, P.E. (2011) Addressing the threat to biodiversity from botanic gardens. *Trends in Ecology & Evolution* 26, 168–174.

Hulme, P.E., Pyšek, P., Jarošík, V., Pergl, J., Schaffner, U. and Vilà, M. (2013) Bias and error in understanding plant invasion impacts. *Trends in Ecology & Evolution* 28, 212–218.

Jeschke, J.M., Keesing, F. and Ostfeld, R.S. (2013) Novel organisms: comparing invasive species, GMOs, and emerging pathogens. *Ambio* 42, 541–548.

Jeschke, J.M., Bacher, S., Blackburn, T.M., Dick, J.T.A., Essl, F., Evans, T., Gaertner, M., Hulme, P.E., Kühn, I., Mrugała, A. *et al.* (2014) Defining the impact of non-native species. *Conservation Biology* 28, 1188–1194.

Kettunen, M., Genovesi, P., Gollasch, S., Pagad, S., Starfinger, U., ten Brink, P. and Shine, C. (2008) Technical support to EU strategy on invasive species (IAS) – Assessment of the impacts of IAS in Europe and the EU (final module report for the European Commission). Institute for European Environmental Policy Brussels, Belgium.

Kolar, C.S. and Lodge, D.M. (2001) Progress in invasion biology: predicting invaders. *Trends in Ecology & Evolution* 16, 199–204.

Kumschick, S., Gaertner, M., Vilà, M., Essl, F., Jeschke, J. M., Pyšek, P., Ricciardi, A., Bacher, S., Blackburn, T.M., Dick, J.T.A. *et al.* (2015) Ecological impacts of alien species: quantification, scope, caveats, and recommendations. *BioScience* 65, 55–63.

Leung, B., Roura-Pascual, N., Bacher, S., Heikkilä, J., Brotons, L., Burgman, M.A., Dehnen-Schmutz, K., Essl, F., Hulme, P.E., Richardson, D.M., Sol, D. and Vilà, M. (2012) TEASIng apart alien species risk assessments: a framework for best practices. *Ecology Letters* 15, 1475–1493.

Lockwood J.L. and McKinney, M.L. (2001) *Biotic Homogenization.* Kluwer Academic/Plenum Publishers, New York.

Lockwood, J.L., Cassey, P. and Blackburnm T. (2005) The role of propagule pressure in explaining species invasions. *Trends in Ecology & Evolution* 20, 223–228.

Lockwood, J.L., Hoopes, M.F. and Marchetti, M.P. (2013) *Invasion Ecology.* 2nd edn. Wiley-Blackwell, Chichester, UK.

Maggi, E., Benedetti-Cecchi, L., Castelli, A., Chatzinikolaou, E., Crowe, T., Ghedini, G., Kotta, J., Lyons, D., Ravaglioli, C., Rilov, G. *et al.* (2015) Ecological impacts of invading seaweeds: a meta-analysis of their effects at

different trophic levels. *Diversity and Distributions* 21, 1–12.

Magurran, A.E. (2016) How ecosystems change. *Science* 351, 448–449.

McLachlan, J.S., Hellmann, J.J. and Schwartz, M.W. (2007) A framework for debate of assisted migration in an era of climate change. *Conservation Biology* 21, 297–302.

Myers, J.H., Simberloff, D., Kuris, A.M. and Carey, J.R. (2000) Eradication revisited: dealing with exotic species. *Trends in Ecology & Evolution* 15, 316–320.

Parker, I.M., Simberloff, D., Lonsdale, W.M., Goodell, K., Wonham, M., Kareiva, P.M., Williamson, M.H., Von Holle, B., Moyle, P.B., Byers, J.E. and Goldwasser, L. (1999) Impact: toward a framework for understanding the ecological effects of invaders. *Biological Invasions* 1, 3–19.

Piaggio, A.J., Segelbacher, G., Seddon, P.J., Alphey, L., Bennett, E.L., Carlson, R.H., Friedman, R.M., Kanavy, D., Phelan, R., Redford, K.H., Rosales, M. *et al.* (2017) Is it time for synthetic biodiversity conservation? *Trends in Ecology & Evolution* 32, 97–107.

Pimentel, D., Lach, L., Zuniga, R. and Morrison, D. (2000) Environmental and economic costs of nonindigenous species in the United States. *BioScience* 50, 53–65.

Ricciardi, A., Hoopes, M.F., Marchetti, M.P. and Lockwood, J.L. (2013) Progress toward understanding the ecological impacts of non-native species. *Ecological Monographs* 83, 263–282.

Ricciardi, A., Blackburn, T.M., Carlton, J.T., Dick, J.T.A., Hulme, P.E., Iacarella, J.C., Jeschke, J.M., Liebhold, A.M., Lockwood, J.L., MacIsaac, H.J. *et al.* (2017) Invasion Science: A horizon scan of emerging challenges and opportunities. *Trends in Ecology & Evolution* 32, 464–474.

Ruiz, G.M. and Carlton, J.T. (2003) Invasion vectors: a conceptual framework for management. In: Ruiz, G.M. and Carlton, J.T. (eds) *Invasive Species: Vectors and Management Strategies.* Island Press, Washington, pp. 459–504.

Sax, D.F. and Brown, J.H. (2000) The paradox of invasion. *Global Ecology and Biogeography* 9, 363–371.

Schlaepfer, M.A., Sax, D.F. and Olden, J.D. (2011) The potential conservation value of non-native species. *Conservation Biology* 25, 428–437.

Seebens, H., Essl, F., Dawson, W., Fuentes, N., Moser, D., Pergl, J., Pyšek, P., van Kleunen, M., Weber, E. and Winter, M. (2015) Global trade will accelerate plant invasions in emerging economies under climate change. *Global Change Biology* 21, 4128–4140.

Seebens, H., Blackburn, T.M., Dyer, E.E., Genovesi, P., Hulme, P. E., Jeschke, J.M., Pagad, S., Pyšek, P., Winter, M., Arianoutsou, M. *et al.* (2017) No saturation in the accumulation of alien species worldwide. *Nature Communications* 8, 14435.

Simberloff, D. (2009a) We can eliminate invasions or live with them. Successful management projects. *Biological Invasions* 11, 149–157.

Simberloff, D. (2009b) The role of propagule pressure in biological invasions. *Annual Review of Ecology, Evolution, and Systematics* 40, 81–102.

Simberloff, D. (2011) Charles Elton: neither founder not siren, but prophet. In: Richardson D.M. and Pyšek, P. (eds) *Fifty Years of Invasion Ecology: the Legacy of Charles Elton.* Blackwell, Oxford, pp. 161–168.

Simberloff, D. and Gibbons, L. (2004) Now you see them, now you don't! – population crashes of established introduced species. *Biological Invasions* 6, 161–172.

Sorte, C.J., Williams, S.L. and Carlton, J.T. (2010) Marine range shifts and species introductions: comparative spread rates and community impacts. *Global Ecology and Biogeography* 19, 303–316.

Strayer, D.L., Eviner, V.T., Jeschke, J.M. and Pace, M.L. (2006) Understanding the long-term effects of species invasions. *Trends in Ecology & Evolution* 21, 645–651.

Strayer, D.L. (2009) Twenty years of zebra mussels: lessons from the mollusk that made headlines. *Frontiers in Ecology and the Environment* 7, 135–141.

Su, S., Cassey, P. and Blackburn, T.M. (2016) The wildlife pet trade as a driver of introduction and establishment in alien birds in Taiwan. *Biological Invasions* 18, 215–229.

Thomsen, M.S., Olden, J.D., Wernberg, T., Griffin, J.N. and Silliman, B.R. (2011a) A broad framework to organize and compare ecological invasion impacts. *Environmental Research* 111, 899–908.

Thomsen, M.S., Wernberg, T., Olden, J.D., Griffin, J.N. and Silliman, B.R. (2011b) A framework to study the context-dependent impacts of marine invasions. *Journal of Experimental Marine Biology and Ecology* 400, 322–327.

Turbelin, A.J., Malamud, B.D. and Francis, R.A. (2017) Mapping the global state of invasive alien species: patterns of invasion and policy responses. *Global Ecology and Biogeography* 26, 78–92.

Vall-llosera, M. and Cassey, P. (2017) Leaky doors: private captivity as a prominent source of bird introductions in Australia. *PloS ONE* 12, e0172851.

Vilà, M., Espinar, J.L., Hejda, M., Hulme, P.E., Jarošík, V., Maron, J.L., Pergl, J., Schaffner, U., Sun, Y. and Pyšek, P. (2011) Ecological impacts of invasive alien plants: a meta-analysis of their effects on species, communities and ecosystems. *Ecology Letters* 14, 702–708.

Williams, F., Eschen, R., Harris, A., Djeddour, D., Pratt, C., Shaw, R., Varia, S., Lamontagne-Godwin, J., Thomas, S.E. and Murphy, S.T. (2010) The economic cost of invasive non-native species on Great Britain. *CABI Proj No VM10066*, 1–99.

Williamson, M. (1993) Invaders, weeds and the risk from genetically manipulated organisms. *Cellular and Molecular Life Sciences* 49, 219–224.

Williamson, M. (1999) Invasions. *Ecography* 22, 5–12.

Wonham, M.J., Walton, W.C., Ruiz, G.M., Frese, A.M. and Galil, B.S. (2001) Going to the source: role of the invasion pathway in determining potential invaders. *Marine Ecology Progress Series* 215, 1–12.

Wonham, M.J., Byers, J.E., Grosholz, E.D. and Leung, B. (2013) Modeling the relationship between propagule pressure and invasion risk to inform policy and management. *Ecological Applications* 23, 1691–1706.

Yokomizo, H., Possingham, H.P., Thomas, M.B. and Buckley, Y.M. (2009) Managing the impact of invasive species: the value of knowing the density–impact curve. *Ecological Applications* 19, 376–386.

2

The Hierarchy-of-hypotheses Approach

Tina Heger[1,2,3]* and Jonathan M. Jeschke[4,5,3]

[1]University of Potsdam, Biodiversity Research/Systematic Botany, Potsdam, Germany; [2]Technical University of Munich, Restoration Ecology, Freising, Germany ; [3]Berlin-Brandenburg Institute of Advanced Biodiversity Research (BBIB), Berlin, Germany; [4]Leibniz-Institute of Freshwater Ecology and Inland Fisheries (IGB), Berlin, Germany; [5]Freie Universität Berlin, Institute of Biology, Berlin, Germany

Abstract

Reviewing empirical evidence for major hypotheses in ecology is connected to two methodological challenges. First, all available empirical tests for a given hypothesis should be identified, collected and organized. This can be difficult due to divergent formulations of the hypothesis in the literature and divergent empirical approaches to address it. Second, the entire body of available data should be used to systematically assess the overall level of support of the hypothesis. Many approaches (e.g. formal meta-analyses) typically use only a fraction of the available data, often because of inconsistencies in the data. In this chapter, we show how the new hierarchy-of-hypotheses (HoH) approach can be used to overcome these challenges.

Introduction

Invasion biology, as Chapter 1, this volume, has shown, is a flourishing research discipline. It has generated a high diversity of major hypotheses and a large quantity of theoretical and empirical studies to test them. But, contrary to what might be expected from philosophical theories about science, the large quantity of published data does not translate directly into an increasing ability to explain and predict biological invasions. As reasons for this phenomenon we can blame, amongst others, the high complexity of invasion processes, their context-dependence, and the significance of socio-cultural influences at each stage of an invasion (Heger *et al.*, 2013). Empirical data are constantly proving that invasion biology's major hypotheses are not capturing the nature of the invasion process well enough: for every single hypothesis in invasion biology, empirical tests can be found that provide positive confirming evidence but there are also many studies that show the opposite (Fridley *et al.*, 2007; Jeschke *et al.*, 2012; Moles *et al.*, 2012; Heger and Jeschke, 2014). The question we would like to address in this book is therefore: which major hypotheses in invasion biology are backed up by empirical evidence, and which should be reformulated or completely discarded?

There are two methodological challenges connected to this question. First, all available empirical tests of a given major invasion hypothesis need to be identified, collected and organized. This task does not

* Corresponding author. E-mail: tina-heger@web.de

seem like a *methodological* challenge at first glance but in fact it is, as outlined below. Second, the major hypothesis in question needs to be systematically confronted with the entire body of available empirical data and scrutinized with respect to its validity and coverage.

The usual process of reviewing how well a certain hypothesis is backed up by evidence is starting with a search for all published studies referring to the hypothesis in question. After sorting through the studies and identifying those that actually report data related to the respective hypothesis – in invasion biology, ecology and many other natural sciences – it usually becomes obvious that it is hard to combine the remaining studies, e.g. in a meta-analysis (Koricheva *et al.*, 2013). Not only will the measured effect sizes differ but the studies will also differ in which *exact question*, that is, which exact formulation of a major hypothesis, they address. The methodological challenge at this point is: how can the results of different empirical studies testing different formulations of a major hypothesis be synthesized?

The Hierarchy-of-hypotheses Approach

Recently, we proposed a tool that can help solve this problem: the hierarchy-of-hypotheses (HoH) approach (Jeschke *et al.*, 2012; Heger *et al.*, 2013; Heger and Jeschke, 2014). Its basic notion is that a major hypothesis can be viewed as an overarching idea on top of a hierarchical system of hypotheses. Major hypotheses are so influential because they cover a large range of cases and are formulated in a general way. Often, they are broad ideas rather than actually testable hypotheses. They have the heuristic purpose of guiding and inspiring theoretical and empirical research. But before they can be tested with an experiment or confronted with observational data, they need to be refined. With the HoH approach, we suggest to make this refinement explicit, to graphically illustrate the result and to use it as a framework for interpreting the results.

An HoH consists of a broad, overarching idea that branches into several more specific formulations of this same idea, and these again branch into more specific formulations and so forth, until a level of refinement is reached that allows for empirical testing of the respective sub-(sub-. . .)hypothesis. Fig. 2.1 gives an impression of what the graphic representation of this process can look like. Here, the broad, overarching idea is: 'the HoH approach fosters synthesis'. This idea includes many other ideas; three of them are depicted here as HI, II and III. Each of the three latter hypotheses can be further refined, which is indicated by the lower-level branches HI.1, HI.2, etc. Depending on the

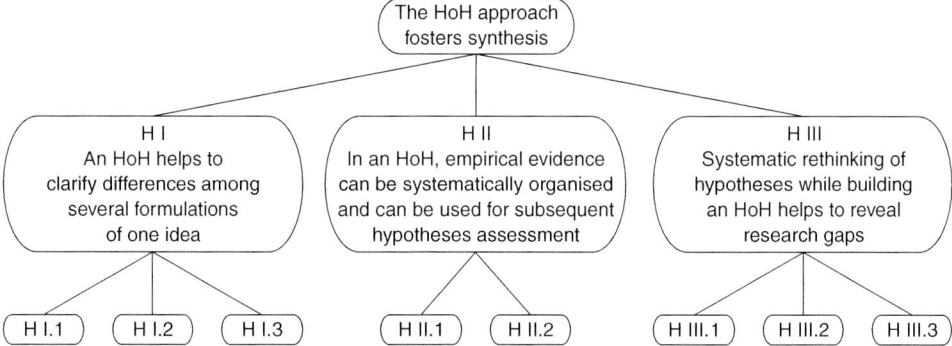

Fig. 2.1. Example of a hierarchy of hypotheses (HoH). A broad, overarching idea is split into more specific formulations, which are further split until a level is reached where the respective hypothesis can be tested with a single empirical study.

area of application, more hierarchical levels can be included.

Assessing Major Hypotheses Using the Hierarchy-of-hypotheses Approach

How does the HoH approach help to meet the named challenges? When the available publications presenting empirical tests of a specific major hypothesis have been identified, e.g. with a systematic review (Moher et al., 2015), they can be organized on the basis of the HoH approach. As a first step, the actual working hypothesis of every study has to be identified. In a second step, all those working hypotheses that relate to the respective major hypothesis can be linked to each other and to the major hypothesis (i.e. the overarching idea).

To make this procedure a bit clearer, we will shortly summarize our analysis from Heger and Jeschke (2014; for an updated version of this study see Chapter 11, this volume). The aim of that study was to assess the validity of the enemy release hypothesis. In its most general formulation, this hypothesis states that 'the absence of enemies in the exotic range is a cause of invasion success'. A systematic literature search with a fixed search term, followed by a screening of full texts of the publications for their relevance, returned a total of 176 empirical tests to be included in the further analysis. We visited the main texts of these publications to analyse in which way the enemy release hypothesis has been tested and which working hypotheses have been addressed. To give an example, Blaney and Kotanen (2001) state in the summary the aim of the study was to test whether 'low rates of attack by natural enemies [. . .] contribute to the invasiveness of exotic plants'. Visiting the main text, it becomes clear that what is investigated in more detail is the following: 'In congeneric pairs of native and exotic plant species, buried seeds of natives are more strongly affected by fungal pathogens than seeds of exotics'. Of the many possibilities to test aspects of the enemy release hypothesis, these authors chose (i) to focus on the performance of species; and (ii) to compare alien to congeneric native species. Other authors instead chose to look at infestation rates or damage (e.g. leaf damage by herbivores) and compared alien species in their native to the same species in the invaded range.

With the help of an HoH, these different approaches to test aspects of the enemy release hypothesis can be disclosed. The overall idea (the absence of enemies in the exotic range is a cause of invasion success) can be sub-divided into sub- and sub-sub-hypotheses according to (i) the chosen indicators for enemy release and (ii) the chosen comparison (see Heger and Jeschke, 2014).

The first methodological challenge for synthesis in invasion ecology identified above was: how can the results of different empirical studies testing a major hypothesis be identified, collected and organized? The example of the enemy release hypothesis shows that the HoH approach makes it possible to clarify the relation of different working hypotheses to an overall idea as well as to each other. It becomes possible to systematically organize available evidence.

The second methodological challenge we identified above was that the respective major hypothesis will have to be confronted systematically with the entire body of available empirical data, and will have to be scrutinized with respect to its validity and coverage. We suggest that the HoH approach is a useful tool to take this challenge as well. In Heger and Jeschke (2014), we exemplified this for the enemy release hypothesis. As a first step, for every empirical test in our database we checked (again using the main text, not only the abstract or summary) whether the working hypothesis we identified in the previous step was supported or questioned by the data, or whether the result was indifferent. For each test, we calculated a weighted score according to the number of species studied and the chosen method (experimental vs observational, field vs laboratory, etc.; Heger and Jeschke, 2014; Chapter 10, this volume). Using these weighted scores, we calculated the level of empirical support for the general idea as well as for every sub- or sub-sub-hypothesis. In this way it can be analysed how well the overall idea is supported by empirical results by taking into account the entire available evidence, and not only those studies that use the same methods and report comparable effect sizes, as is done in a typical meta-analysis. Additionally, the coverage of the hypothesis becomes explicit and can be analysed. In the case of the enemy release hypothesis, it became clear that some sub-hypotheses are receiving ample support, whereas others are largely questioned by empirical tests.

An open question is how much empirical support can be considered to be enough to keep a certain hypothesis and to integrate it into the body of ecological theory, and how much questioning evidence should lead to the refusal of a hypothesis. There seems to be consent that in ecology a single 'yes' or 'no' empirical result does not decide on the usefulness of a hypothesis (Scheiner, 2013; Henriksson et al., 2015). It is still unclear, though, which level of support is necessary. The community needs to discuss this question and we cannot answer it here. For Heger and Jeschke (2014), as well as for this book, we preliminarily suggest that if more than 50% of the weighted evidence is in favour of a major hypothesis, it should not be disregarded. Of course, it should be kept in mind that the level of empirical support depends on the domain within which the hypothesis is tested, e.g. the habitat or taxonomic group (e.g. Henriksson et al., 2015). We suggest that if a major hypothesis is questioned by more than 50% of the empirical tests, it should be reformulated. The HoH approach will be useful for such a reformulation because it helps to reveal which formulations of a given idea exist and how well they are empirically supported. If reformulation does not help to reduce questioning empirical support below 50%, we suggest to disregard the hypothesis as a major hypothesis because the prediction derived from a *major* hypothesis should be more reliable than a coin toss within a *major* domain. The hypothesis might still be kept as a specialized hypothesis for a small domain if it reaches more empirical support there.

Again, the scientific community needs to discuss this issue further. We invite readers to keep in mind that, depending on, for example, your philosophical mind-set, other thresholds than 50%, particularly higher ones, could be equally suitable as the low threshold of 50% suggested here. An alternative approach would be a relative one, which does not use any threshold but instead considers those hypotheses as useful that receive higher empirical support in a certain discipline, compared to the other hypotheses in the same discipline.

Other Applications

The HoH approach can not only be used for the synthesis of empirical data and the assessment of existing hypotheses. It can also be used as a tool to structure ideas before starting empirical analyses and to systematically develop testable hypotheses from theory. To give an example, it is also possible to develop an HoH for the enemy release hypothesis starting from theoretical considerations. Possible sub-hypotheses could be: 'only for species strongly regulated by enemies in the native range, the absence of enemies in the exotic range is a cause of invasion success', or 'only if the invading species show ecological traits that differ from traits of native prey will they suffer from less predation'. On the basis of these sub-hypotheses, a range of working hypotheses could be developed.

In a similar way, it is possible to structure whole research programmes, e.g. in the preparation of a grant proposal for an extensive research project involving several groups. The overall idea for the project in this case forms the top of the hierarchy. Three to five hypotheses that go into a bit more detail can be located at a lower level and these are split further. Every sub-project within the research group links its working hypotheses to a lower level hypothesis. In this way it is possible to conceptually link the different working hypotheses to the overarching idea and make the relation of the different sub-projects to each other explicit. Also, the graphical display can contribute to clarity in communication within the research group.

From a methodological point of view, we believe the HoH approach could be used to advance invasion ecology as well as ecology in general, and it can also be extended to other natural sciences, even to the humanities and arts. In an HoH, disparate fields of research addressing one overall idea or topic can be linked and therefore synthesized. There is a tendency for more and more specialized research, which makes synthesis even more difficult (Sergio et al., 2014); the HoH approach could be increasingly valuable in bridging specialist research results.

The approach also offers a possibility to conceptually and graphically link general ideas to context-specific hypotheses, and thus to mirror complexity. Complexity, context-dependence and socio-cultural influences are challenges that have to be faced when describing, explaining and predicting ecological systems in our globally changing world. Evans and colleagues (2013) point out that in ecology prediction needs to be detached from generality. With respect to predictions based on ecological modelling, the authors call for a combination of complex, system-specific models and simple models. We suggest adding the HoH approach to the ecological toolbox as an additional method to approach complexity and context dependence. Other disciplines facing similar challenges, e.g. the social sciences, could similarly benefit from the approach. Thus, this book is a first test whether the HoH approach is useful as a conceptual backbone for a scientific subdiscipline. If it proves useful for invasion biology, we hope to see it applied in other disciplines as well.

References

Blaney, C.S. and Kotanen, P.M. (2001) Effects of fungal pathogens on seeds of native and exotic plants: a test using congeneric pairs. *Journal of Applied Ecology* 38, 1135–1147.

Evans, M.R., Grimm, V., Johst, K., Knuuttila, T., de Langhe, R., Lessells, C.M., Merz, M., O'Malley, M.A., Orzack, S.H., Weisberg, M. *et al.* (2013) Do simple models lead to generality in ecology? *Trends in Ecology & Evolution* 28, 578–583.

Fridley, J.D., Stachowicz, J.J., Naeem, S., Sax, D.F., Seabloom, E.W., Smith, M.D., Stohlgren, T.J., Tilman, D. and Von Holle, B. (2007) The invasion paradox: Reconciling pattern and process in species invasions. *Ecology* 88, 3–17.

Heger, T. and Jeschke, J.M. (2014) The enemy release hypothesis as a hierarchy of hypotheses. *Oikos* 123, 741–750.

Heger, T., Pahl, A.T., Botta-Dukat, Z., Gherardi, F., Hoppe, C., Hoste, I., Jax, K., Lindström, L., Boets, P., Haider, S. *et al.* (2013) Conceptual frameworks and methods for advancing invasion ecology. *Ambio* 42, 527–540.

Henriksson, A., Yu, J., Wardle, D.A. and Englund, G. (2015) Biotic resistance in freshwater fish communities: species richness, saturation or species identity? *Oikos* 124, 1058–1064.

Jeschke, J.M., Gómez Aparicio, L., Haider, S., Heger, T., Lortie, C.J., Pyšek, P. and Strayer, D.L. (2012) Support for major hypotheses in invasion biology is uneven and declining. *NeoBiota* 14, 1–20.

Koricheva, J., Gurevitch, J. and Mengerson, K. (eds) (2013) *Handbook of Meta-analysis in Ecology and Evolution.* Princeton University Press, Princeton, New Jersey.

Moher, D., Shamseer, L., Clarke, M., Ghersi, D., Liberati, A., Petticrew, M., Shekelle, P., Stewart, L. and PRISMA-P Group (2015) Preferred reporting items for systematic review and meta-analysis protocols (PRISMA-P) 2015 statement. *Systematic Reviews* 4, 1–9.

Moles, A.T., Flores-Moreno, H., Bonser, S.P., Warton, D.I., Helm, A., Warman, L., Eldridge, D.J., Jurado, E., Hemmings, F.A., Reich, P.B. *et al.* (2012) Invasions: the trail behind, the path ahead, and a test of a disturbing idea. *Journal of Ecology* 100, 116–127.

Scheiner, S.M. (2013) The ecological literature, an idea-free distribution. *Ecology Letters* 16, 1421–1423.

Sergio, F., Schmitz, O.J., Krebs, C.J., Holt, R.D., Heithaus, M.R., Wirsing, A.J., Ripple, W.J., Ritchie, E., Ainley, D., Oro, D. *et al.* (2014) Towards a cohesive, holistic view of top predation: a definition, synthesis and perspective. *Oikos* 123, 1234–1243.

3

Hierarchy of Hypotheses or Hierarchy of Predictions? Clarifying Key Concepts in Ecological Research

Alejandro G. Farji-Brener[1]* and Sabrina Amador-Vargas[2]

[1]Lab Ecotono, INIBIOMA, Universidad Nacional del Comahue and CONICET, Bariloche, Argentina; [2]Smithsonian Tropical Research Institute, Balboa, Panamá

Abstract

Hypotheses and predictions are the core of scientific research, and are fundamentally different. Nonetheless, predictions are constantly confused with hypotheses in the scientific literature. The hypothetic-deductive method starts with a question to which potential explanations – hypotheses – are proposed. Each hypothesis has expected outcomes – predictions – that are deduced assuming the hypothesis was true. In this chapter, we point out that the hierarchy-of-hypotheses (HoH) approach is confusing the terms hypothesis and prediction, and further explain why the so-called sub-hypotheses of the HoH approach are in fact predictions. We also propose an alternative approach inspired by the HoH approach to help arrange predictions by their potential to disprove the hypothesis.

Ideas can be Tested only through their Consequences

Hypotheses are potential explanations to observed patterns. When a baby is crying, how can you discriminate among the hypotheses that it's hungry, cold or that it has pooped? Because hypotheses are abstractions, their validity can only be tested indirectly: deducing their theoretical consequences and contrasting them with the data. Hence, ideas and their consequences (i.e. expected results or predictions) are intrinsically different concepts. In ecology, hypotheses are 'intellectual gambles' about how nature works, whereas predictions are the expected outcomes assuming that hypotheses are true. Predictions are derived (i.e. deduced) from hypotheses but it is unlikely to deduce a hypothesis from a prediction. In practical terms, it should be easy to differentiate hypotheses from predictions; hypotheses should be expressed by phrasing, whereas predictions should be expressed by plotting. In other words, if what you are calling a 'hypothesis' can be plotted in an x-y graph, then it is not a hypothesis but a prediction derived from an – unspecified – hypothesis. Despite this difference, ecologists often formulate predictions but erroneously state them as hypotheses (Farji-Brener, 2003). This confusion is not trivial because it violates several principles of the hypothetic-deductive method, generating

* Corresponding author. E-mail: alefarji@yahoo.com

© CAB International 2018. *Invasion Biology: Hypotheses and Evidence*
(eds J.M. Jeschke and T. Heger)

conceptual and methodological misunderstandings. First, formulating predictions without explicit mention of the hypothesis they are derived from inhibits readers to judge the deductive capacity of the author. Second, omitting explicit hypotheses in the text (or replacing them with expected outcomes) complicates the interpretation of the results because readers are actually ignoring the tested idea. Consequently, it is impossible to assess whether relevant predictions for a particular hypothesis were evaluated or whether non-mentioned alternative hypotheses are also valid explanation for the phenomena studied. We believe that the hierarchy-of-hypotheses (HoH) approach, as it was originally formulated, exemplifies how the confusion between hypotheses and predictions could overshadow an interesting framework to study biological invasions (Farji-Brener and Amador-Vargas, 2014). Here we point out this misunderstanding and propose an alternative method that better differentiates explanations from their expected consequences.

About Hierarchies, Hypotheses and Predictions in the HoH Approach

Hierarchy can be understood as any system of things ranked one above another; however, the placement of a subject in the ranking depends on the criteria under which the hierarchical system was built. For example, the criteria may be the spatial scale (e.g. large to small), a temporal sequence (e.g. past to present) or responsibility level (e.g. supervisor to employee), among others. The HoH approach uses the criteria of scope. The HoH is arranged with a central hypothesis that 'cover[s] a large range of cases and [is] formulated in a general way' on top of a hierarchical system of more specific, low-ranked hypotheses (Heger and Jeschke, 2014; and Chapter 2, this volume). We found two weaknesses in this method. First, a well-defined hypothesis needs to be specific. Explicit potential explanations generate precise predictions that can be subject to tests but general explanations are often impossible to assess because it is problematic

to deduce their consequences. Thus, hypotheses formulated in a 'general way' are often not useful to design research. Second, and probably more important, arranging 'hypotheses' in a theoretical hierarchical system encourages the already common confusion between hypotheses and predictions and ways to organize information. For example, as previously discussed in Farji-Brener and Amador-Vargas (2014), the classification of sub-hypotheses in Heger et al. (2013) and Heger and Jeschke (2014) to evaluate the enemy release hypothesis (ERH) seems to be a tool for organizing published results rather than a set of hypotheses. The three categories used by the authors to classify the sub-hypotheses (indicator of enemy release, type of comparison and type of enemies) are not explanations derived from the ERH of why invaders are successful. The first and second classification criteria focus on the response variable (e.g. leaf damage, performance of alien species, abundance or diversity of specialist or generalist natural enemies, etc.) and the last criterion focuses on the treatments used in the experimental design (e.g. comparison between alien vs native species, with or without enemies, etc.). In fact, the so-called 'sub-hypotheses' are really predictions (less damage by enemies, less infestation by enemies and enhanced performance; Fig. 1 in Heger et al., 2013; Heger and Jeschke, 2014) because they are not new explanations but expected results assuming the ERH is true. The second-order level of 'sub-hypotheses' specifies the treatments or units of comparison used in those predictions (leaf damage by enemies, alien vs natives, alien in native vs exotic range) but again, those are not different or independent explanations, as different hypotheses should be. Following the hypothetic-deductive method, in the case of biological invasions the question is: what is the cause of invasion success? In the example of Heger and Jeschke (2014; and Chapter 11, this volume) a proposed explanation is the enemy release hypothesis: 'the absence of enemies in the exotic range', and the 176 empirical tests that were evaluated are in fact specific predictions (expected results) of that single hypothesis and are not sets of specific sub-hypotheses of the ERH. Recognizing this confusion between hypotheses and predictions shows that the field of biological invasions actually lacks an overwhelming number of hypotheses, although it has an astounding set of predictions and results. This fact might encourage the evaluation of new, creative and independent explanations to the question of invasion success.

An Alternative: the Hierarchies-of-Expected-Outcomes Approach (HoEO)

As explained before, the validity of a hypothesis can only be tested through its consequences (i.e. predictions). However – and unlike hypotheses – predictions can be easily arranged in a hierarchy because they can differ in their capacity to discard erroneous ideas. The strongest predictions (i.e. key assumptions) are those with the best potential to discard the hypothesis they have been derived from; hence, they can be formulated following the hypothesis at the first branch level. Other, more specific predictions may be arranged at lower levels (Fig. 3.1a). This hierarchy-of-expected-outcomes approach (HoEO) follows the general principles of the HoH approach but uses consequences of hypotheses instead of sub-hypotheses, offering some conceptual and logistical

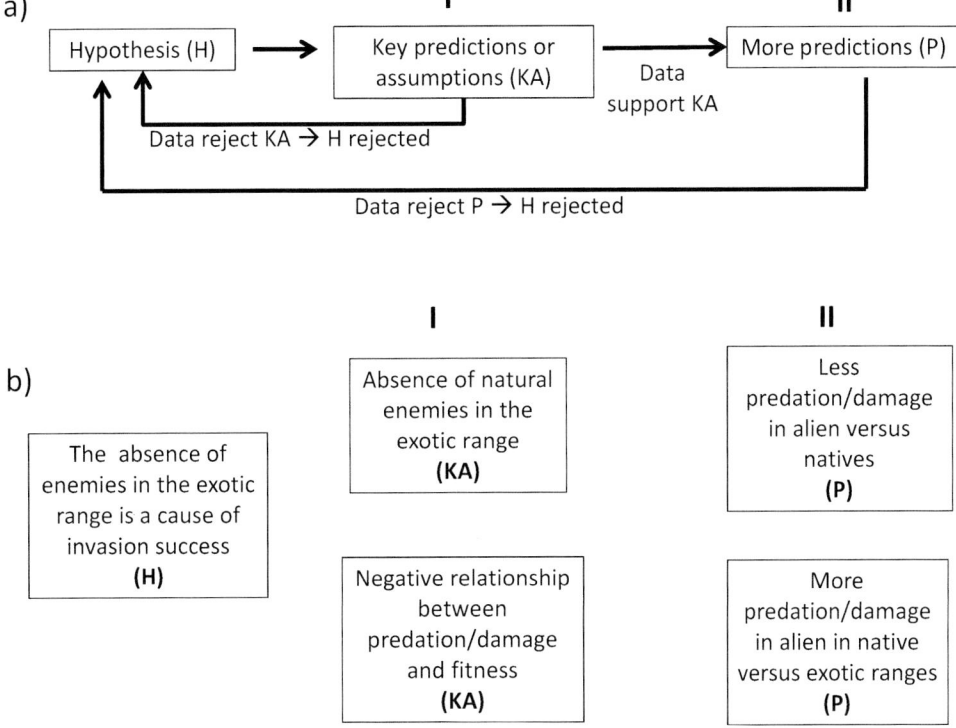

Fig. 3.1. A schematic representation of (a) the hierarchy-of-expected-outcomes approach (HoEO) and (b) an example of its potential application. I and II are hierarchical levels in decreasing order. The HoEO is arranged with a central hypothesis (H) on top of a hierarchical system of their expected outcomes assuming the hypothesis is true. The expected outcomes are arranged according to their potential to discard the hypothesis they have been derived from; first key predictions or assumptions (KA) and at lower levels more specific predictions (P). If the data fail to support the key assumptions, the hypothesis is rejected and it is worthless to invest energy and time testing second-order expected outcomes. For example, if enemies are present in the exotic range of an invasive species, the 'absence' of enemies cannot be a cause of invasion success and no further testing for that hypothesis would be needed.

advantages. First, this method clearly follows the hypothetic-deductive method, which states that ideas can be tested only through their consequences. Second, it helps to clearly distinguish hypotheses from predictions, improving the logistics and organization of data collection. Third, it forces researchers to know the natural history of the studied system or organism to formulate the pertinent predictions. Fourth, the HoEO avoids extra work: it is worthless to invest energy and time testing second-order expected outcomes when data did not support key assumptions. For example, if damage by natural enemies does not reduce fitness or if natural enemies are present in the exotic range, the 'absence' of enemies cannot be a cause of invasion success and no further testing is needed (Fig. 3.1b). Conversely, if key assumptions are supported by data, we can proceed to test specific predictions, which might also branch in lower levels of hierarchy, relevance or peculiarity depending on the topic. This last point might help in the debate about how much support does a hypothesis need. In fact, the hypothetic-deductive method works by trying to disprove a hypothesis; hence, finding a result that agrees with a prediction of the ERH should not be considered as evidence proving that hypothesis right, but a result that does not agree with a prediction of the ERH is in fact evidence against the hypothesis (Popper, 1959). Confirming assumptions do not necessarily validate the proposed hypothesis but are an essential step before continuing with the next level of expected outcomes. Finally, this conceptual framework allows contrasting alternative hypotheses more clearly. A priori assignation of different weights to the expected outcomes within each hypothesis may help discriminate which hypothesis explains a natural pattern better. The system for weighting predictions and the decision of which hypothesis has more support could follow the method proposed by Heger and Jeschke (2014; and Chapter 2, this volume) to weight 'sub-hypotheses'.

We agree with Heger and Jeschke (Chapter 2, this volume) that relevant hypotheses need to be identified and systematically confronted with the available empirical data to understand a natural phenomenon. But from our point of view, the proposal that a major hypothesis can be viewed as an overarching idea on top of a hierarchical system of more specific hypotheses could be a source of confusion rather than of clarity. As previously discussed, this approach may promote the incorrect formulation of expected outcomes as ideas. Conversely, expected outcomes can be easily arranged in a hierarchical system because they naturally differ in their strength to overthrow the idea from which they came. As originally proposed in the HoH approach, the HoEO method also shows other useful applications in a broad field of scientific disciplines. It can be used as a tool to structure ideas before starting empirical analyses, to analytically develop testable hypotheses from theory, to structure whole research programmes and to graphically clarify concepts. In summary, the hierarchy-of-outcomes approach could be a simple, practical, and useful tool to better understand how nature works.

References

Farji-Brener, A.G. (2003) Uso correcto, parcial e incorrecto de los términos hipótesis y predicciones en ecología. *Ecología Austral* 13, 223–227.

Farji-Brener, A. G. and Amador-Vargas, S. (2014) Hierarchy of hypotheses or cascade of predictions? A comment on Heger *et al.* (2013) *Ambio* 43, 1112–1114.

Heger, T. and Jeschke, J.M. (2014) The enemy release hypothesis as a hierarchy of hypotheses. *Oikos* 123, 741–750.

Heger, T., Pahl, A.T., Botta-Dukat, Z., Gherardi, F., Hoppe, C., Hoste, I., Jax, K., Lindström, L., Boets, P., Haider, S. *et al.* (2013) Conceptual frameworks and methods for advancing invasion ecology. *Ambio* 42, 527–540.

Popper, K. (1959) *The Logic of Scientific Discovery.* Basic Books, New York.

4

Mapping Theoretical and Evidential Landscapes in Ecological Science: Levins' Virtue Trade-off and the Hierarchy-of-hypotheses Approach

James Griesemer[1,2]*

[1]University of California, Davis, Philosophy, Davis, California, USA;
[2]University of California, Davis, Center for Population Biology,
Davis, California, USA

Abstract

According to Richard Levins, there is no single, best all-purpose model in ecology, nor can useful models simultaneously maximize theoretical virtues of generality, realism and precision (Levins, 1966, 1968a,b). A 'virtue trade-off' must be made if models are to be useful for understanding, explanation, prediction or control. Recently, large-scale data synthesis projects have aimed to systematically review many empirical studies to evaluate major hypotheses. A specific version of systematic review called the 'hierarchy-of-hypotheses' approach has been proposed (Jeschke *et al.*, 2012; Heger and Jeschke, 2014; and Chapter 2, this volume). I argue that investigation of the patchwork of models organized by Levins' theoretical 'virtue space' and discovery of robust theorems across a *theoretical* landscape could be enhanced by recent efforts to manage evidential complexity in a patchwork of empirical studies by the hierarchy-of-hypotheses approach.

Introduction

Richard Levins' well-known argument for theoretical pluralism, that there is no single, best all-purpose model in ecology, was bolstered by his equally famous conjecture that useful models could not simultaneously maximize theoretical virtues of generality, realism and precision (Levins, 1966, 1968a,b). A virtue trade-off must be faced if models are to be manageable, and thus useful, for understanding, explanation, prediction and control. It has been argued that Levins' pluralism about models is a pragmatic response to idealistic research programmes that sought to represent all ecological complexity in a single unified or overarching comprehensive ecosystem model, which might be conceivable but which, Levins argued, would be unmanageable, uninterpretable and largely untestable (Odenbaugh, 2006). Recently, large-scale data synthesis projects have aimed to collect data from many empirical studies interpreted as bearing on a single major hypothesis or several related hypotheses. These include quantitative meta-analyses organized by statistical measures of effect sizes and qualitative systematic literature reviews bearing on major hypotheses. Recently, a specific version of systematic review called the 'hierarchy-of-hypotheses' (HoH) approach has been proposed (Jeschke *et al.*, 2012; Heger and Jeschke, 2014; and this

* Corresponding author. E-mail: jrgriesemer@ucdavis.edu

volume). The organization of ecological complexity in a patchwork of models according to a theoretical 'virtue space' and discovery of robust theorems across a theoretical landscape, I argue, is complemented by recent efforts to manage evidential complexity in a patchwork of empirical studies according to a dataset virtue space organizing a hierarchy of empirical hypotheses.

Ecological Theory and Systematic Reviews of Evidence

If Levins is right, theory in ecology is better represented by a patchwork or network of models than by a compact set of general laws (Levins, 1966, 1968a,b). Fundamental principles there may be but the set of theoretical models in ecology that expresses them does not look much like fundamental theories in quantum or relativity physics. Models in ecology, Levins said, face trade-offs of generality, realism, precision and manageability. He famously conjectured that at most two of the first three could be simultaneously maximized in any one model. The desire to maximize all three leads to unmanageable models that we can imagine but not use or make sense of to any practical effect. In the 1960s, the introduction of large-scale computing and computer simulation into science fuelled some desires for a systems ecology that could solve Levins' conjecture by letting computers manage what was for humans unmanageable. It is fair to say that Levins' picture of theory in ecology (1968b) turned out closer to right than, say, Kenneth Watt's (1968).

A new desire for systematic approaches is emerging with the rise of big data and online access to digital data. Systematic review of empirical datasets and their use in evaluation of general empirical hypotheses has become popular owing to easy electronic search access to large numbers of datasets stored in online journals and databases. Meta-analysis is perhaps the most familiar type of systematic review in the sciences. Meta-analysis seeks to combine empirical studies measuring the same effect, typically using statistics to combine studies with varying sample and effect sizes and to test expectations about effect size quantitatively. More broadly, systematic review is any assembly of data or empirical results in order to bring them to bear as qualitative or quantitative evidence salient to a general claim. If meta-analysis is an 'evidence synthesis' methodology seeking to characterize how evidence is dispersed around a particular hypothesis 'peak', e.g. by calculating weighted average effect size, then systematic review more broadly could be interpreted as an 'evidence landscape' methodology, seeking to characterize the distribution of evidence over the landscape of support for a particular hypothesis.

The HoH approach in ecology uses systematic review of data and findings from the accessible empirical literature to test general hypotheses. In this commentary, I focus on Heger and Jeschke's use of the approach to study hypotheses about biological invasion but the argument should apply more generally wherever Levins' picture of theory and trade-off conjecture holds (Jeschke et al., 2012; Heger and Jeschke, 2014; and this volume). The main goal of this commentary is to interpret the HoH approach in terms of some philosophical ideas about scientific models and to use Levins' conception of theory in ecology and his conjecture about maximizing 'theoretical virtues' (generality, realism and precision) of ecological models as a means to pose new theoretical and philosophical questions about systematic review and 'big data' driven research. My argument is that choices of how to structure a hierarchy of hypotheses for use in evaluation of data as empirical evidence of a general hypothesis involve theoretical as well as empirical trade-offs and that consideration of systematic empirical reviews can inform theory development as well as empirical practice.

Locating empirical studies included in an HoH investigation in an *empirical* virtues space of dataset and study design trade-offs, and mapping them into Levins' *theoretical* virtues space of modelling trade-offs can point to new research questions for ecologists using systematic review methods. It may also suggest new avenues for philosophers seeking to understand the 21st century shift of attention from theory-driven to data-driven science (Leonelli, 2016). New philosophical research questions raised by the HoH approach concern the relation between empirical and theoretical science in practice that should also be of great interest to ecologists thinking about the broad direction of their field. Specifically, I propose that the HoH approach may be useful for exploring the efficiency and effectiveness of the current distribution of empirical research effort to address general hypotheses on topics such as biological invasions as well as the robustness of empirical results to the idealizing assumptions of the theoretical models that ground empirical hypotheses.

An HoH analysis proceeds as follows (Heger and Jeschke, 2014; and Chapter 2 this volume). A broad, *overarching hypothesis* is re-expressed in refined, specific formulations of *working hypotheses*, and these in turn are re-expressed in *operational hypotheses* allowing for direct empirical testing. Choices in the concepts, modes of refinement and deployment of a variety of empirical research strategies by many different investigators lead to a hierarchy of hypotheses actually investigated with variety at each level of refinement. Heger and Jeschke (2014), for example, evaluated 176 empirical studies of the overarching 'enemy release' hypothesis to explain the success of invasive biological species (see also Chapter 11, this volume). The overarching hypothesis is that release from enemies in their native range is a cause of success of invading species in an alien range. The broad hypothesis generalizes over a wide range of enemy release mechanisms by using the word 'cause' to refer to them all, by abstract reference to 'species' and generically to native and alien ranges. This cause is articulated in working hypotheses concerning specific mechanisms or forms of causation (Fig. 1 in Heger and Jeschke, 2014). The 'difference-making' of effects by these mechanisms can be realized, and thus measured and used as evidence to test operational hypotheses, in a variety of ways.

The general hypothesis is refined, making it more precise or more realistic by identifying a range of indicators of mechanisms for release of alien invading species from predation, such as reduced damage or infestation or enhanced performance in the new range. Specific contrasts (between particular alien and native species, particular alien populations in native and invaded ranges, particular invasive and non-invasive aliens) serve in a second level of sub-hypotheses that can be empirically tested. Empirical support for these lowest-level hypotheses in the HoH approach is accepted according to whatever criteria were used in the original studies, avoiding a limitation of quantitative meta-analysis in that there need not be consistency of statistical methodology but only a qualitative conclusion regarding the *direction* of support with respect to each sub-hypothesis: positive, negative or neutral. Support for hypotheses at higher levels in the hierarchy is aggregated from support at the lower levels, according to criteria imposed by the systematic review.

One advantage of the HoH approach over quantitative meta-analysis is that HoH does not require any common measure of effect size or even a shared notion of the effect to be explained to include empirical studies in a systematic review and HoH test of a general hypothesis. But the price of relaxing the constraint to a shared effect in favour of the more abstract assessment of direction of support is that theoretical, conceptual work is required to articulate both a general hypothesis and specific, empirically testable versions of it in terms of component concepts, which must be further articulated in a hierarchy of empirically feasible and operational studies. In a sense, that conceptual work, together with the empirical choices needed to reach the operational level, introduce a degree of subjectivity into systematic review.

Levins was concerned to explore the robustness of empirical findings to the array of idealizing assumptions – 'biases' in a technical sense (see Wimsatt, 2007) – in models trading off one theoretical virtue against others to make models manageable. Robustness analysis is a way of recovering objectivity from subjective or biased studies: if individual studies depend on subjective, idealizing assumptions, results spanning many models or studies are robust to the extent to which they do not depend on those idealizing assumptions. 'Our truth', Levins famously said, 'is the intersection of independent lies' (Levins, 1966, p. 423). My interest in HoH is to explore the robustness of empirical tests using systematic review methodology to idealizing assumptions over the array of practical trade-offs employed to make empirical work manageable, such as the choice of how many native and alien species to include in a study and how to study them. I propose that this robustness property can be explored by considering how the empirical studies of HoH analyses are distributed across Levins' space of theoretical virtues. A further extension of this mapping project would include the consideration of values (constitutive and contextual; Longino, 1990) that also constrain both modelling and empirical virtues trade-offs.

Models and Hypotheses

A preliminary step is needed to connect models to hypotheses so that the HoH approach to evaluate general empirical hypotheses using systematic review of datasets can be linked to consideration of theoretical models. Levins' account of theoretical virtue trade-offs between generality, realism, precision and manageability concerned models rather than hypotheses. What some philosophers have had to say about the models–hypotheses connection turns out to be quite useful in this context.

Models and hypotheses are linked in an account of evidence evaluation by the philosopher Ronald Giere (1988, 1997). On Giere's account, a theoretical model identifies a kind of 'structure' – a way some part of the physical world *might be*. A theoretical hypothesis links a theoretical model to the world by *asserting* that the model fits that part of the physical world in some respects, to some degree of accuracy. (Here I let Giere's invocation of accuracy stand in for any theoretical virtue, including all of Levins'.) The assertion is true if the model fits in the

specified respects and degrees, and false otherwise. The project of gathering evidence by observation, measurement or experiment from that part of the physical world and bringing it to bear on the truth or falsity of a hypothesis involves reasoning from models to predictions about what data *should* look like, if the model were a good fit. Thus, prediction-data comparisons yield evaluations of model-world fit and thus assessment of the truth/falsity of hypotheses.

The salience of Giere's picture of evidence evaluation is that for every hypothesis for which truth is in question, there is a model for which fit is in question. When HoH constructs a hierarchy of hypotheses, a corresponding hierarchy of models is implied, even if models cannot be directly read off the hypotheses or conversely. Nevertheless, we can ask Levins' questions about the implied models of the hierarchy of hypotheses: do they trade-off generality for realism and precision, realism for generality and precision, or precision for generality and realism? If there is no trade-off evident, is the model unmanageably complex? Whether trade-offs are evident in the model or not, would empirical tests of associated hypotheses be empirically manageable?

Hierarchy-of-Hypotheses Analysis

To conduct an HoH systematic review, a fixed-term keyword search of a digital database of empirical studies is used to identify, collect and organize available empirical tests of a given hypothesis. The database could simply be the Internet itself or a more limited, systematically constructed archive of empirical research papers in online accessible journals or archiving services such as Web of Science, PubMed or JSTOR. The selection of keywords requires concept work that relates the overarching hypothesis to more specific, empirically testable forms so as to generate a collection of empirical papers suitable for organization in an HoH. I assume without argument here that the choice and articulation of the concepts used to conduct the keyword search reflect or guide theoretical virtue trade-offs in models implied by the hypotheses recovered from the literature. Whether the concepts reflect or guide, in turn, depends on whether models derive from efforts to represent the structure of empirically investigated phenomena or from efforts to translate first principles into possible structures (see Griesemer, 2013, on the distinction between empirical and theoretical modelling approaches).

The primary data serving as evidence regarding a general hypothesis such as the enemy release hypothesis in invasion ecology is, for each empirical study, whether the data supported, questioned or was indifferent regarding the truth of the hypothesis. But in order to aggregate support into a summary evaluation over the evidential landscape, some criterion for aggregation is needed. Various criteria might be considered, e.g. >50% of studies in one direction (positive, negative) of support provide aggregate support at that level of the hierarchy. Heger and Jeschke (2014) constructed a more sophisticated weighted score for each study before aggregating, based on number of species studied and the chosen method of investigation (experimental vs observational, field vs enclosure vs laboratory; see Heger and Jeschke, 2014; and Chapter 2, this volume). Using the weighted scores, they calculated level of empirical support for the general hypothesis as well as for every working and operational hypothesis. Weighting criteria, I suggest, reflect a complex mix of theoretical, empirical and practical virtue considerations that are often entwined with additional value considerations. For example, more species studied in a given empirical investigation might suggest greater generality of results, yet fewer might be investigated in any given study on grounds of time, expense or potential harm to species due to the study methods. Some methods of study suggest trade-offs of greater realism vs greater precision (e.g. observing more species in the field vs experimenting with fewer species in the laboratory). Heger and Jeschke's weighting function in terms of number of species considered and method of investigation might

indicate preferences in generality/precision trade-offs.

These difficult conceptual issues cannot be sorted out here. In the remainder of the commentary, I explore research questions for ecology and for philosophy that depend on sorting out how evidential weighting schemes might reflect preferences for theoretical virtues in 'Levins trade-offs' and preferences for empirical virtues in 'Heger–Jeschke trade-offs' and how these might in turn be used to guide the design of future empirical research and theoretical modelling.

Heger and Jeschke (2014, p. 746) raise a number of questions based on their HoH analysis of the enemy release hypothesis:

> Our findings also pose several questions: Why do studies more frequently show a reduced infestation of alien species than they show an increase in performance? Why do alien species frequently seem to be released from enemies in their invaded as compared to their native range, but less often as compared to native species? Why is the enemy release hypothesis frequently supported by studies on vertebrates but rarely by studies on plants? And why is there more supporting evidence in marine than in terrestrial systems? When addressing these questions, future studies might reveal important insights into the ecology of biological invasions.

These questions all concern the empirical phenomenon of biological invasion and enemy release and how future studies might target unanswered empirical questions. I suggest parallel theoretical and philosophical questions arise from the HoH approach about patterns of ecological research. If we were able to locate each of the 176 empirical studies Heger and Jeschke analyse in Levins' space of theoretical trade-offs, we might be able to articulate these theoretical questions more precisely.

The hypotheses at different levels of an HoH reflect different trade-offs of generality, realism and precision, and to different extents in the diverse empirical studies. By implication, these trade-offs may be reflected in predictions from models or may even guide how models are constructed in the first place. The overarching hypothesis in an HoH analysis is general by design. Working hypotheses sacrifice some generality for the sake of realism by substituting a specific type of mechanism hypothesized to be operating in real systems. Operational hypotheses trade away some realism in favour of empirical manageability of the study, e.g. by considering a small number of species in native or alien ranges (thus excluding some species interactions from study); or they trade away precision in the study, e.g. by using observational methods yielding a cruder measure of cause or effect than can be achieved in the laboratory; or they trade away generality by studying a limited number of species in a restricted range of habitats, e.g. terrestrial vertebrates only, or a single plant–insect interaction. It might turn out, then, that when Heger and Jeschke weight empirical studies according to criteria such as number of species studied and method of investigation, they produce a scheme that could be useful for locating hypotheses and models in Levins' space of *theoretical* virtues and trade-offs corresponding to the empirical trade-offs and virtues of operational studies and datasets.

Linking Theoretical and Empirical Virtue Spaces

Suppose we could classify each implied model at every level of an HoH analysis in terms of the generality, realism and precision with which it purports to 'fit'/be true of the world. Abstractly, Levins' claim that generality, realism and precision cannot simultaneously be maximized simply means that one corner of the three-dimensional 'Levins' space' will be unoccupied and that there will be various 'surfaces' of manageable models that jointly maximize these virtues under the trade-off constraints (see also Matthewson and Weisberg, 2009). But where do the empirical studies of an HoH fall in Levins' space? Just as Heger and Jeschke ask why empirical studies seem to show that the enemy release hypothesis is frequently supported by studies on

vertebrates but rarely by studies on plants, we might ask of an HoH systematic review whether studies investigating particular hypotheses/models at any of the three levels of the hierarchy – overarching, working or operational – cluster in some portion of Levins' theoretical virtue space. Do studies supporting enemy release tend to investigate hypotheses clustering in, say, the high realism–precision/low generality region of the space, while studies questioning enemy release cluster in, say, high generality–precision/low realism regions? Or perhaps studies supporting some working hypotheses regarding particular mechanisms of enemy release tend to occur in some regions, whereas studies questioning that mechanism cluster in another region.

Suppose it turned out that those studies supporting enemy release more frequently do so via a reduced infestation of alien species than via an increase in performance, and at the same time they tend to be framed in terms of hypotheses/models that cluster in the 'low realism' part of Levins' space. A possible explanation is that precise measures of degree of damage do not actually reflect the extent of release from enemies but crude measures of damage coupled with high realism in representing how damage interacts with performance can reflect the extent of enemy release. Such a finding might suggest that before concluding, on the basis of the HoH evidence, that reduced infestation rather than increased performance is a major mechanism, studies based on hypotheses/models with greater realism should be performed. Differently put: are low-realism studies robust to the idealizing theoretical assumption that sacrifices realism for generality and precision?

Consider the *unweighted* results of each empirical study as positive, negative or inconclusive evidence (+/−/0) supporting an operational, working and overarching hypothesis in an HoH analysis. Suppose we use *weighting* functions, such as the one Heger and Jeschke (2014, eq. 1, p. 743) produce, not as support weights in HoH analysis but as general models to quantify Levins' space, e.g. number of species studied reflects degree of 'realism'. Modes of investigation, e.g. lab vs field, reflects ranges of 'precision'. Taxonomic scope reflects degree of generality. Then suppose we plot each empirical study in an HoH analysis as three connected points in Levins' space corresponding to the operational, working and overarching models the study is used to test. We could treat each connected trio as a directed graph of support as in the HoH, from operational to working to overarching hypothesis/model. Colour the paths according to whether they provide positive, negative or neutral support. It might be of interest to know whether an even distribution of studies (points) across Levins' space results in more (or less) robust results at a given hierarchical level of an HoH than collections of studies concentrated in small regions of Levins' space. The mapping could provide a measure for the efficiency of current ecological studies to assess overarching hypotheses, i.e. whether preferences for certain theoretical virtues and trade-off choices, which become manifest in patterns of empirical work, tend to limit or enhance the power of systematic reviews to provide evidence regarding general hypotheses. Perhaps any particular theoretical virtue in a given model/hypothesis is not so important provided empirical studies sacrifice it differently, e.g. by each studying a different limited range of taxa so that the collection of studies in an HoH are robust to sacrificing generality; or perhaps realism turns out to be an extremely sensitive virtue to degree of taxonomic generality because including many rather than few species in studies of limited taxonomic scope better reflects the complex dynamics of species interactions governing enemy release mechanisms and the number of species–species interactions drops geometrically with arithmetic drop in number of species included in a study.

Finally, the discovery of robust or non-robust patterns of support for models making different trade-offs in theoretical virtues may reveal 'aesthetic' preferences of modellers for certain theoretical virtues that do not serve the empirical community very well. Such discoveries may help refocus theoretical efforts on classes of models that have a better chance of meeting that highly desired desideratum: falsifiability.

References

Giere, R. (1988) *Explaining Science: a Cognitive Approach.* University of Chicago Press, Chicago, Illinois.

Giere, R. (1997) *Understanding Scientific Reasoning,* 4th edn. Harcourt, Brace, Jovanovich, Fort Worth, Texas.

Griesemer, J. (2013) Formalization and the meaning of 'theory' in the inexact biological sciences. *Biological Theory* 7, 298–310.

Heger, T. and Jeschke, J.M. (2014) The enemy release hypothesis as a hierarchy of hypotheses. *Oikos* 123, 741–750.

Jeschke, J.M., Gómez Aparicio, L., Haider, S., Heger, T., Lortie, C.J., Pyšek, P. and Strayer, D.L. (2012) Support for major hypotheses in invasion biology is uneven and declining. *NeoBiota* 14, 1–20.

Leonelli, S. (2016) *Data-centric Biology: a Philosophical Study.* University of Chicago Press, Chicago, Illinois.

Levins, R. (1966) The strategy of model building in population biology. *American Scientist* 54, 421–431.

Levins, R. (1968a) *Evolution in Changing Environments; Some Theoretical Explorations. Monographs in Population Biology no. 2.* Princeton University Press, Princeton, New Jersey.

Levins, R. (1968b) Ecological engineering: theory and technology. *Quarterly Review of Biology* 43, 301–305.

Longino, H. (1990) *Science as Social Knowledge: Values and Objectivity in Scientific Inquiry.* Princeton University Press, Princeton, New Jersey.

Matthewson, J. and Weisberg, M. (2009) The structure of tradeoffs in model building. *Synthese* 170, 169–190.

Odenbaugh, J. (2006) The strategy of 'The strategy of model building in population biology'. *Biology and Philosophy* 21, 607–621.

Watt, K. (1968) *Ecology and Resource Management: a Quantitative Approach.* McGraw-Hill, New York.

Wimsatt, W.C. (2007) *Re-engineering Philosophy for Limited Beings: Piecewise Approximations to Reality.* Harvard University Press, Cambridge, Massachusetts.

5

A Hierarchy of Hypotheses or a Network of Models

Samuel M. Scheiner[1]* and Gordon A. Fox[2]

[1]*US National Science Foundation, Alexandria, Virginia, USA;*
[2]*University of South Florida, Tampa, Florida, USA*

Abstract

Many different explanations for biological invasions have been proposed, yet there is little evidence that any of them are general. Invasion biologists are increasingly favouring a multicausal outlook rather than seeking single explanations of invasions, or seeking to identify a list of characteristics of invasive species. The time is ripe for systematization of ideas in invasion biology. The hierarchy-of-hypotheses (HoH) framework proposed by Jeschke and Heger thus comes at an opportune moment but we believe that it must be modified substantially to be useful. A principal problem is the notion that science proceeds simply by testing hypotheses, which neglects the roles of models (the source of hypotheses) and of the broader constitutive theories that give rise to models. Accepting or rejecting hypotheses seems to us a doomed approach for complex phenomena like invasions, in which multiple causes are likely to account for the differences between invading and native populations. This multicausality also means that explanations for invasions are not likely to be strictly hierarchical. Finally, the criteria proposed for retaining hypotheses in the HoH – support by a majority of empirical studies – is ad hoc and beset by severe statistical problems. We argue that examination of a network of models – rather than a hierarchy of hypotheses – is needed to move the field forward.

Introduction

> Prediction is very difficult, especially about the future.
>
> Danish proverb

Jonathan Jeschke and Tina Heger have put forward a framework for exploring and testing a set of related hypotheses that relies on arranging those hypotheses into a hierarchy; they then illustrate that concept by examining theories of species invasions (see Chapter 2, this volume). We provide a critique of both the general approach and the specific application, noting its many valuable aspects while suggesting ways in which it needs improvement. We do so by contrasting their approach with ones that we have developed for generally relating sets of theories (Scheiner and Willig, 2011) and specifically organizing invasion models (Gurevitch *et al.*, 2011). The framework proposed by Jeschke and Heger raises both scientific and philosophical issues. Both sets of issues converge around ideas of how to organize our thinking about scientific theories and how to carry out empirical research about invasions because the core problems are deeply linked: how do we understand complex multicausal phenomena such as invasions?

From an early focus on identifying characteristics of invasive species (e.g. wide physiological tolerances, rapid growth, early reproduction; Drake *et al.*, 1989; Gilpin, 1990), the attention of most invasion biologists shifted to a continually growing list of

* Corresponding author. E-mail: sscheine@nsf.gov

named ecological hypotheses intended to explain why organisms might be invasive pests in a new region but not where they are native. Catford *et al.* (2009) identified 29 named hypotheses, including the well-known enemy release hypothesis and evolution of increased competitive ability hypothesis. Although each hypothesis is plausible in principle, most invasion studies have been observational, simply asking whether the assumptions or predicted pattern of a given hypothesis matched those in a particular system. Few studies tested multiple hypotheses (Lowry *et al.*, 2013), fewer have examined the possibility of multicausation (Gurevitch *et al.*, 2011; Catford *et al.*, 2016) and very few, if any, have estimated the magnitudes of multiple causes or examined their interactions. No hypothesis has proved particularly well supported or general (Jeschke *et al.*, 2012).

If we have not been very successful at assembling lists of characteristics of invasive species nor at identifying a core set of mechanisms that promote invasions, what strategies should researchers consider in trying to understand, predict and manage biological invasions? The following strategies suggest themselves:

- Stay the course. Perhaps we have simply not yet identified a sufficient set of hypotheses (or, for that matter, a sufficient set of species characteristics).
- Shift the focus from species characteristics to site characteristics (Davies *et al.*, 2005; Fridley *et al.*, 2007).
- Focus on interactions between species characteristics and site characteristics (Perkins and Nowak, 2013).
- Eschew mechanisms and make purely statistical predictions (e.g. Williamson and Fitter, 1996; Kolar and Lodge, 2001; Marchetti *et al.*, 2004; Simberloff, 2009).
- Focus on mechanisms, especially multicausal pathways (Gurevitch *et al.*, 2011; Catford *et al.*, 2016).
- Abandon hope.

These strategies are not mutually exclusive – more than one might be useful – and it is not an exhaustive list.

Hypotheses versus Models

Choosing among these strategies can be helped by considering how scientific knowledge is accumulated and organized. Theories can be considered as hierarchical frameworks that connect broad general principles to highly specific models (Scheiner and Willig, 2011). Scheiner and Willig termed families of models constitutive theories. A given constitutive theory delimits a domain within which constituent models are interconnected so as to form a coherent entity that focuses on one or a few phenomena in need of explanation. A model is a specific type of theory that represents or simplifies the natural world. From models come hypotheses, which are testable statements derived from or representing various components of the model (Pickett *et al.*, 2007). Jeschke and Heger take a different approach that they dubbed a hierarchy of hypotheses (HoH; see Chapter 1, this volume). The underlying notion is that there are overarching forms of hypotheses (e.g. invasions may be facilitated by the fact that natural enemies are missing in the invaded region) and more narrow versions of such hypotheses (e.g. buried seeds of native species are more strongly affected by fungal pathogens than congeneric exotics). A better name for their approach, however, would be a hierarchy of theories or a hierarchy of models. Although these might sound like simple differences in word choice, theories and models are not synonymous with hypotheses. Jeschke and Heger make clear that in their hierarchy the higher levels are generalizations and any tests occur at the lowest level. If so, hypotheses only exist at that lowest level, not across the entire hierarchy.

We make this distinction between hypothesis and model because we emphasize that the family of models within a constitutive theory need not be, and probably is not, a simple hierarchy. Rather, a constitutive theory can contain multiple, non-exclusive models or processes. Two models might be attempts to explain the same phenomenon but have different structures. For example, one model might be spatially explicit and the other not. One model might attempt to explain the entire domain of the theory, whereas a related model might confine itself to a portion of that domain. In the case of species invasions, a model explaining all parts of the theory would be quite complex (see Fig. 2 in Gurevitch *et al.*, 2011), whereas a model of invasion via predator release would be considerably simpler (see Fig. 3a in Gurevitch *et al.*, 2011). By recognizing that theories contain families of models, it is possible to build new models by assembling pieces of existing models. A number of examples are discussed in Catford *et al.* (2009) and Gurevitch *et al.* (2011). The result is a network of interconnected models. Some models might be related hierarchically in the sense that one model is a subset of another but there is no necessity for such a relationship. Contrast the Jeschke and Heger hierarchy with the process network of Gurevitch *et al.* (2011) (Fig. 5.1). The latter

framework allows for the mixing and matching of components as needed for any given system.

Models are used for different purposes. In ecology, those purposes can roughly be divided into understanding versus prediction and management. Of course, a given instance might partake of both usages. These two usages manifest in the study of species invasion. The process of invasion includes many phenomena that we wish to better understand, such as long-distance dispersal, competitive exclusion, niche relations, extinction and co-evolution (Chesson, 2000; Sakai *et al.*, 2001; Shea and Chesson, 2002; Webb, 2003; Hastings *et al.*, 2005; Melbourne *et al.*, 2007; MacDougall *et al.*, 2009; Nuismer *et al.*, 2010; Bolnick *et al.*, 2011; Kelly *et al.*, 2011; Kubisch *et al.*, 2014). At the same time, invasive species are of concern both for the management of natural

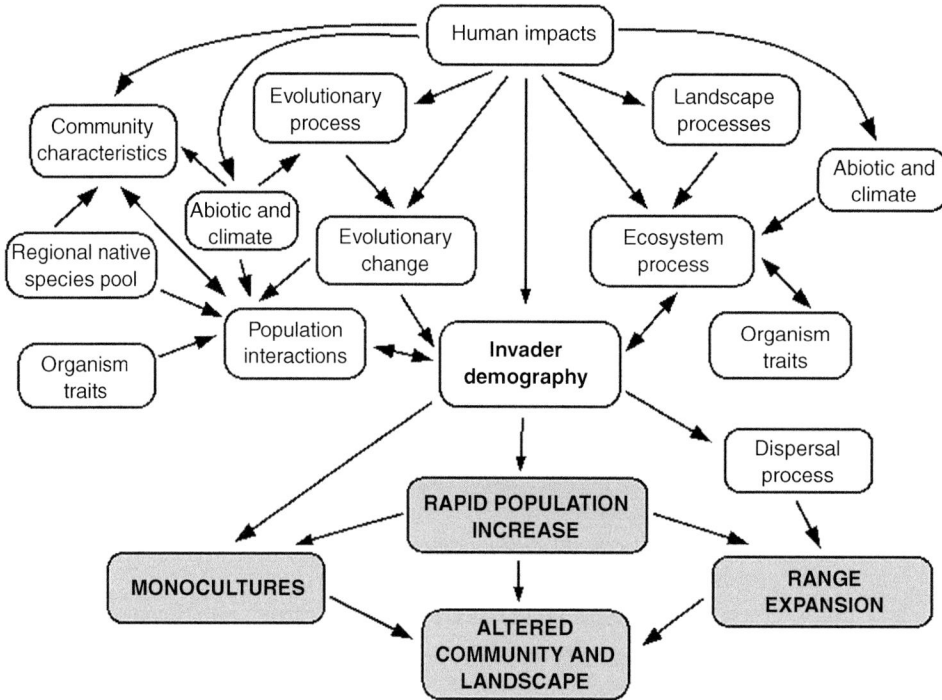

Fig. 5.1. Causal network for biological invasions. Four different aspects of invasions (dark boxes) are shown. Invasion outcomes are mediated, at least partly, by invader demography. Each box can be further expanded to describe additional processes. All of the named invasion hypotheses – such as enemy release or evolution of increased competitive ability – are examples of particular processes (and only those processes) operating to produce an invasion. (From Gurevitch *et al.*, 2011.)

areas and the minimization of economic harm (Pimentel, 2001). If we desire understanding of general phenomena, bringing together a broad family of models can be illuminating by allowing for connections among processes that we may not have considered. Invasion biology, however, has been plagued by the lack of a general theory and general models. Instead, considerable attention has been devoted to considering particular narrow hypotheses (such as enemy release) and to multiplying the number of named hypotheses. One hopeful sign is the recent development of general frameworks that can integrate many of the ideas in the plethora of invasion hypotheses (Catford et al., 2009; Gurevitch et al., 2011). These frameworks can be viewed as the beginnings of a general theory.

Conversely, if we need to manage a specific invasion, we want to pare our model to only those processes that are relevant to that situation. Adding irrelevant components can cost time and resources for gathering unneeded data and can obscure policy decisions. That paring, however, should start from the broadest starting point because it can be even worse to omit relevant processes than to include irrelevant ones. In this context, too, it is generally a poor strategy to begin by asking whether individual processes should or should not be included in a model.

In ecology, when we attempt to gain understanding we typically ask a question of the form: what is the relative importance of these various processes for this phenomenon? For example, in species invasions we may ask how often invasions involve the evolution of new capabilities after a species has reached a new area versus winnowing among pre-existing phenotypes (Sakai et al., 2001). Maron et al. (2004) studied the importance of contemporary selection, founder effects and phenotypic plasticity in the spread within North America of the European native Hypericum perforatum and found evidence for contemporary adaptation to the climate in the introduced range. This study was a test of those processes, a hypothesis test, only in the sense of asking, 'did post-invasion evolution occur in

H. perforatum?' Their answer of 'Yes' does not tell us that such a process is general. After all, we know that evolution occurs, and the process can occur in any system in which the invader is genetically variable. However, if we look for this process in many systems, we might conclude that it is rare. In fact, post-invasion adaptation of invaders appears to be rather common (Siemann and Rogers, 2001; Lee, 2002; Blair and Wolfe, 2004; Bossdorf et al., 2005; Keller and Taylor, 2008) but not universal (Willis et al., 2000).

There are very few examples in ecology of ever showing that a hypothesized process is completely wrong or absent. One of the few instances was Clements' (1937) superorganism theory that was completely abandoned because it was built on an incorrect understanding of evolutionary processes. More often theories get reworked. For example, Hubbell's (2001) neutral theory has been reformulated as a quasi-neutral theory (Alonso et al., 2006) because of concerns about the assumptions of ecological equivalence among species and because measured patterns did not match predictions (e.g. Fargione et al., 2003; Paul et al., 2009). Or a process may be restricted to a narrow domain or set of conditions. For example, competitive exclusion occurs quite easily under the classic Lotka–Volterra model but subsequent theory has greatly restricted the conditions under which it occurs (Chesson, 2000).

Testing Invasion Processes

Jeschke and Heger present their HoH structure as performing two functions: organizing hypotheses and then testing them (see Chapter 2, this volume). First, what is being organized is a set of proposed processes. By their own description:

> Major hypotheses are so influential because they cover a large range of cases and are formulated in a general way. Often, they are broad ideas rather than actually testable hypotheses. They have the heuristic purpose of guiding and inspiring theoretical and empirical research. (Chapter 2, this volume)

Substituting 'theory' and 'model' for 'hypotheses' in the first and second sentence, respectively, would make their formulation entirely consistent with our approach. But that is a matter of words.

More critical is their second function, hypothesis testing. Although their presentation sounds as if they want to measure the relative importance of a given process, that is not what they do:

> An open question is how much empirical support can be considered to be enough to keep a certain hypothesis and to integrate it into the body of ecological theory, and how much questioning evidence should lead to the refusal of a hypothesis . . . For Heger and Jeschke (2014), as well as for this book, we preliminarily suggest that if more than 50% of the weighted evidence is in favour of a major hypothesis, it should not be disregarded . . . If reformulation does not help to reach an empirical support above 50%, we suggest to disregard the hypothesis as a major hypothesis because the prediction derived from a *major* hypothesis should be more reliable than a coin toss within a *major* domain. The hypothesis might still be kept as a specialized hypothesis for a small domain if it reaches more empirical support there. (Chapter 2, this volume)

This formulation fails both with regard to understanding and to prediction and management. How often something occurs is not related to its domain. Mutation is a rare event when measured one organism at a time, yet it is applicable to all life. There are many other such processes in ecology and evolution, including long-distance dispersal, survival of juveniles in many species and the occurrence of major disturbances. Each of these are uncommon or rare events that are nevertheless quite important. Suggesting that a process represents just a specialized hypothesis because it is infrequently observed might lead managers to ignore that process during initial assessment.

It can be quite misleading to study invasions simply by focusing on hypothesis testing. To see this, begin by asking what invasion endpoint is being evaluated (Gurevitch *et al.*, 2011). One might examine invasive species in terms of demographic processes (e.g. survival), population growth rate, rate of spatial spread, total population size or geographic area covered, community dominance, or changes in the rate of nitrogen cycling; this list is by no means exhaustive. Each of these quantities may measure something related to invasions and they are not necessarily correlated with one another. Evaluation of, say, the effect of predator escape might lead to very different conclusions for these different quantities; there are many different natural enemies hypotheses and they are not mutually exclusive.

Nor is there reason to think that the same processes operate at all times during an invasion (Dietz and Edwards, 2006). If different processes operate at different times, then a strategy based on simple hypothesis tests is likely to fail. Multiple processes are involved in many, if not most, invasions. Kelly *et al.* (2011), reviewing over 10,000 records of plant species naturalizations around the world, concluded that the distribution of naturalizations was consistent with the view that naturalizations are idiosyncratic. This does not imply that, say, natural enemies may not play important roles in invasions but rather that none of the sorts of mechanisms described in the 29 hypotheses summarized by Catford *et al.* (2009) are likely to be general explanations for invasions.

There are also statistical reasons to be sceptical about simple null hypothesis testing as a general approach to studying the causes of invasions. One reason is that it is likely that the great majority of species introductions do not lead to invasion and are never studied (Williamson and Fitter, 1996; Holt *et al.*, 2005). The literature is quite unlikely to represent a random sample of introductions or even of successful invasions.

Indeed, it seems clear that a more useful strategy for understanding the processes involved in any particular invasion would involve considering multiple processes and their interaction. Unfortunately, the Jeschke and Heger division of hypotheses into sub-hypotheses and sub-sub-hypotheses fails to recognize that processes can have non-additive effects. Consider the difficulty of studying a system with an introduced predator. In a simple scenario, one might expect predator population growth to depend on prey availability. Among the reasons more predators may not result in fewer prey are: (i) compensatory mortality in the prey population; (ii) predators concentrating on prey classes with low reproductive value; (iii) a prey population regulated by factors other than predation; and (iv) a predator functional response that leads to non-linear predator–prey dynamics. Such interactions mean that any attempt to aggregate results of individual studies as an overall test of a process might be misleading.

Beyond the severe problems already discussed, there are deep difficulties with the procedure Jeschke and Heger propose to weight evidence and combine the results of multiple studies. First, the weighting scheme described in Heger and Jeschke (2014) is arbitrary. There is no statistical justification for weighting experimental field studies as being eight times more valuable than observational studies, for example. Using the square root of the number of species as an indication of sample size treats species as sampling units, ignoring sample sizes within species. Second, once this weighting has been performed, what remains in the procedure suggested by Jeschke and Heger is precisely a vote count. The argument Jeschke and Heger use – that meta-analysis requires that studies use the same methods and comparable effect sizes (see Chapter 2, this volume) – is incorrect (Koricheva et al., 2013; Gurevitch and Nakagawa, 2015). Like any statistical method, meta-analysis can be done poorly or inappropriately. There are certainly times one cannot (or should not) perform a meta-analysis. But the approach suggested by Jeschke and Heger is not the answer.

Conclusion

Invasions are multicausal and idiosyncratic. This does not imply that there are no useful empirical generalizations to be made about them. Rather, it suggests the importance of multicausal models, both in understanding particular invasions and in synthesizing the results of multiple studies. As more systems are examined, the relative importance of various processes can be assessed. The network framework of Gurevitch et al. (2011) can then be used to summarize that assessment in an easily conveyed format by altering the widths of the arrows (Fig. 5.1), while making sure that additional information makes clear under what conditions various sets of processes are likely to be present. Such a network of models provides a rich starting point for both understanding and management. Ecological processes are not strictly hierarchical and neither should be research projects nor management plans.

Acknowledgements

GAF was partly supported by grant DEB-1120330 from the US National Science Foundation. This manuscript is based on work done while SMS was serving at the US National Science Foundation. The views expressed in this paper do not necessarily reflect those of the National Science Foundation or the United States Government.

References

Alonso, D., Etienne, R.S. and McKane, A.J. (2006) The merits of neutral theory. *Trends in Ecology & Evolution* 21, 451–457.

Blair, A.C. and Wolfe, L.M. (2004) The evolution of an invasive plant: an experimental study with *Silene latifolia*. *Ecology* 85, 3035–3042.

Bolnick, D.I., Amarasekare, P., Araújo, M.S., Bürger, R., Levine, J. M., Novak, M., Rudolf, V.H.W., Schreiber, S.J., Urban, M.C. and Vasseur, D. A. (2011) Why intraspecific trait variation matters in community ecology. *Trends in Ecology & Evolution* 26, 183–192.

Bossdorf, O., Auge, H., Lafuma, L., Rogers, W., Siemann, E. and Prati, D. (2005) Phenotypic and genetic differentiation between native and introduced plant populations. *Oecologia* 144, 1–11.

Catford, J. A., Jansson, R. and Nilsson, C. (2009) Reducing redundancy in invasion ecology by integrating hypotheses into a single theoretical framework. *Diversity and Distributions* 15, 22–40.

Catford, J. A., Baumgartner, J. B., Vesk, P. A., White, M., Buckley, Y. M. and McCarthy, M. A. (2016) Disentangling the four demographic dimensions of species invasiveness. *Journal of Ecology* 104, 1745–1758.

Chesson, P. (2000) Mechanisms of maintenance of species diversity. *Annual Review of Ecology and Systematics* 31, 343–366.

Clements, F.E. (1937) Nature and structure of the climax. *Journal of Ecology* 24, 252–284.

Davies, K.F., Chesson, P., Harrison, S., Inouye, B.D., Melbourne, B.A. and Rice, K.J. (2005) Spatial heterogeneity explains the scale dependence of the native-exotic diversity relationship. *Ecology* 86, 1602–1610.

Dietz, H. and Edwards, P.J. (2006) Recognition that causal processes change during plant invasion helps explain conflicts in evidence. *Ecology* 87, 1359–1367.

Drake, J.A., Mooney, H.A., di Castri, F., Groves, R.H., Kruger, F.J., Rejmanek, M. and Williamson, G.M. (eds) (1989) *Biological Invasions: a Global Perspective* – SCOPE 37. Wiley, Chichester, UK.

Fargione, J., Brown, C.S. and Tilman, D. (2003) Community assembly and invasion: An experimental test of neutral versus niche processes. *Proceedings of the National Academy of Sciences USA* 100, 8916–8920.

Fridley, J.D., Stachowicz, J.J., Naeem, S., Sax, D.F., Seabloom, E.W., Smith, M.D., Stohlgren, T.J., Tilman, D. and Holle, B.V. (2007) The invation paradox: reconciling pattern and process in species invations. *Ecology* 88, 3–17.

Gilpin, M. (1990) Biological invasions. *Science* 248, 88–89.

Gurevitch, J. and Nakagawa, S. (2015) Research synthesis methods in ecology. In: Fox, G.A., Negrete-Yankelevich, S. and Sosa, V.J. (eds) *Ecological Statistics: Contemporary Theory and Application*. Oxford University Press, Oxford, UK, pp. 200–227.

Gurevitch, J., Fox, G.A., Wardle, G.M., Inderjit and Taub, D. (2011) Emergent insights from the synthesis of conceptual frameworks for biological invasions. *Ecology Letters* 14, 407–418.

Hastings, A., Cuddington, K., Davies, K.F., Dugaw, C.J., Elmendorf, S., Freestone, A., Harrison, S., Holland, M., Lambrinos, J., Malvadkar, U. *et al.* (2005) The spatial spread of invasions: new developments in theory and evidence. *Ecology Letters* 8, 91–101.

Heger, T. and Jeschke, J.M. (2014) The enemy release hypothesis as a hierarchy of hypotheses. *Oikos* 123, 741–750.

Holt, R.D., Barfield, M. and Gomulkiewicz, R. (2005) Theories of niche conservatism and evolution: could exotic species be potential tests. In: Sax, D.F., Stochowicz, J.J. and Gaines, S.D. (eds) *Species Invasions: Insights into Ecology, Evolution, and Biogeography*. Sinauer, Sunderland, Massachusetts, pp. 259–290.

Hubbell, S.P. (2001) *The Unified Neutral Theory of Biodiversity and Biogeography*. Princeton University Press, Princeton, New Jersey.

Jeschke, J.M., Gómez Aparicio, L., Haider, S., Heger, T., Lortie, C.J., Pyšek, P. and Strayer, D.L. (2012) Support for major hypotheses in invasion biology is uneven and declining. *Neo-Biota* 14, 1–20.

Keller, S.R. and Taylor, D.R. (2008) History, chance and adaptation during biological invasion: separating stochastic phenotypic evolution from response to selection. *Ecology Letters* 11, 852–866.

Kelly, C.K., Blundell, S.J., Bowler, M.G., Fox, G.A., Harvey, P.H., Lomas, M.R. and Ian Woodward, F. (2011) The statistical mechanics of community assembly and species distribution. *New Phytologist* 191, 819–827.

Kolar, C.S. and Lodge, D.M. (2001) Progress in invasion biology: predicting invaders. *Trends in Ecology & Evolution* 16, 199–204.

Koricheva, J., Gurevitch, J. and Mengerson, K. (eds) (2013) *Handbook of Meta-analysis in Ecology and Evolution*. Princeton University Press, Princeton, New Jersey.

Kubisch, A., Holt, R.D., Poethke, H.-J. and Fronhofer, E.A. (2014) Where am I and why? Synthesizing range biology and the eco-evolutionary dynamics of dispersal. *Oikos* 123, 5–22.

Lee, C.E. (2002) Evolutionary genetics of invasive species. *Trends in Ecology & Evolution* 17(8), 386–391.

Lowry, E., Rollinson, E.J., Laybourn, A.J., Scott, T.E., Aiello-Lammens, M.E., Gray, S.M., Mickley, J. and Gurevitch, J. (2013) Biological invasions: a field synopsis, systematic review, and database of the literature. *Ecology and Evolution* 3, 182–196.

MacDougall, A.S., Gilbert, B. and Levine, J.M. (2009) Plant invasions and the niche. *Journal of Ecology* 97, 609–615.

Marchetti, M.P., Moyle, P.B. and Levine, R. (2004) Invasive species profiling? Exploring the characteristics of non-native fishes across invasion stages in California. *Freshwater Biology* 49, 646–661.

Maron, J.L., Vilà, M., Bommarco, R., Elmendorf, S. and Beardsley, P. (2004) Rapid evolution of an invasive plant. *Ecological Monographs* 74, 261–280.

Melbourne, B.A., Cornell, H.V., Davies, K.F., Dugaw, C.J., Elmendorf, S., Freestone, A.L., Hall, R.J., Harrison, S., Hastings, A., Holland, M. *et al.* (2007) Invasion in a heterogeneous world: resistance, coexistence or hostile takeover? *Ecology Letters* 10, 77–94.

Nuismer, S.L., Gomulkiewicz, R. and Ridenhour, B.J. (2010) When is correlation coevolution? *American Naturalist* 175, 525–537.

Paul, J.R., Morton, C., Taylor, C.M. and Tonsor, S.J. (2009) Evolutionary time for dispersal limits the extent but not the occupancy of species' potential ranges in the tropical plant genus *Psychotria* (Rubiaceae). *American Naturalist* 173, 188–199.

Perkins, L.B. and Nowak, R.S. (2013) Invasion syndromes: hypotheses on relationships among invasive species attributes and characteristics of invaded sites. *Journal of Arid Land* 5, 275–283.

Pickett, S.T.A., Kolasa, J. and Jones, C.G. (2007) *Ecological Understanding: the Nature of Theory and the Theory of Nature,* 2nd edn. Elsevier, New York.

Pimentel, D. (ed.) (2001) *Biological Invasions,* 2nd edn. CRC Press, Boca Raton, Florida.

Sakai, A.K., Allendorf, F.W., Holt, J.S., Lodge, D.M., Molofsky, J., With, K.A., Baughman, S., Cabin, R.J., Cohen, J.E., Ellstrand, N.C. *et al.* (2001) The population biology of invasive species. *Annual Review of Ecology and Systematics* 32, 305–332.

Scheiner, S.M. and Willig, M.R. (2011) A general theory of ecology. In: Scheiner, S.M. and Willig, M.R. (eds) *The Theory of Ecology.* University of Chicago Press, Chicago, Illinois, pp. 3–19.

Shea, K. and Chesson, P. (2002) Community ecology theory as a framework for biological invasions. *Trends in Ecology & Evolution* 17, 170–176.

Siemann, E. and Rogers, W. E. (2001) Genetic differences in growth of an invasive tree species. *Ecology Letters,* 4, 514–518.

Simberloff, D. (2009) The role of propagule pressure in biological invasions. *Annual Review of Ecology, Evolution, and Systematics* 40, 81–102.

Webb, C. (2003) A complete classification of Darwinian extinction in ecological interactions. *American Naturalist* 161, 181–205.

Williamson, M. and Fitter, A. (1996) The varying success of invaders. *Ecology* 77, 1661–1666.

Willis, A.J., Memmott, J. and Forrester, R.I. (2000) Is there evidence for the post-invasion evolution of increased size among invasive plant species? *Ecology Letters* 3, 275–283.

6

The Hierarchy-of-hypotheses Approach Updated – a Toolbox for Structuring and Analysing Theory, Research and Evidence

Tina Heger[1,2,3]* and Jonathan M. Jeschke[4,5,3]

[1]University of Potsdam, Biodiversity Research/Systematic Botany, Potsdam, Germany; [2]Technical University of Munich, Restoration Ecology, Freising, Germany; [3]Berlin-Brandenburg Institute of Advanced Biodiversity Research (BBIB), Berlin, Germany; [4]Leibniz-Institute of Freshwater Ecology and Inland Fisheries (IGB), Berlin, Germany; [5]Freie Universität Berlin, Institute of Biology, Berlin, Germany

Abstract

In Chapters 3, 4 and 5, three groups of authors have commented on the hierarchy-of-hypotheses (HoH) approach as introduced in Chapter 2. Here, we make suggestions on how to account for the issues raised in Chapters 3–5, whilst also considering several comments by other colleagues that we received during discussions or in response to our presentations on this topic. We focus on five issues and address each in a separate section of this chapter. In the concluding section, we suggest treating the HoH approach as a toolbox; we describe its core as well as an array of modules that can be chosen.

Hypotheses – or Predictions – or Models?

In Chapter 3, this volume, Alejandro Farji-Brener and Sabrina Amador-Vargas ask what the hierarchy-of-hypotheses (HoH) approach actually deals with: hypotheses or rather predictions. Sam Scheiner and Gordon Fox in Chapter 5 argue that, in fact, the approach should be named hierarchy of theories or hierarchy of models rather than hierarchy of hypotheses, and Farji-Brener and Amador-Vargas suggest a hierarchy of expected outcomes.

In ecology, and also in the philosophy of science, the terms 'predictions', 'hypotheses', 'models' and 'theories' are not used consistently. When developing the approach, we had a broad definition in mind, with hypotheses not necessarily being 'specific' (Farji-Brener and Amador-Vargas, Chapter 3) or restricted to 'testable statements derived from or representing various components of [a] model' (Scheiner and Fox, Chapter 5). We agree that such statements are hypotheses, but our understanding of 'hypotheses' is broader. We use this term in a way that major, overarching ideas are included as well. This view is backed, for example, by the rather broad definition of Giere (1988, 1997; cited in Griesemer, Chapter 4, this volume; see also Murray, 2004). Also, we think that it is often hard to draw a clear line between hypotheses in the sense of 'new explanations' and predictions in the sense of

* Corresponding author. E-mail: tina-heger@web.de

'expected outcomes' (Farji-Brener and Amador-Vargas, Chapter 3). For example, the 'key assumptions' in the hierarchy of expected outcomes (Fig. 3.1b, Chapter 3, e.g. 'absence of natural enemies in the exotic range') in our opinion could be interpreted as predictions but also as hypotheses.

To be more explicit about the different hierarchical levels in our approach, we would like to follow the suggestion in Chapter 4 by James Griesemer and make the following distinctions:

- *Overarching hypotheses* are the major, broad ideas at the top of the hierarchy, as introduced in Chapter 2;
- *Working hypotheses* are the refinements of these broad overarching ideas at the following, lower levels; and
- *Operational hypotheses* allow for direct empirical testing.

Working hypotheses in many cases can be developed from an overarching hypothesis on theoretical grounds, whereas operational hypotheses will have to additionally consider practical constraints such as the need to apply a certain *research approach* (which response variables are measured; which comparison is made?) or to focus on a specific *study system*. We think these terminological differentiations can help avoid misunderstanding, and we will test their usefulness in Chapter 11 for developing a revised version of an HoH for the enemy release hypothesis.

Should Studies be Weighted According to their Methods and, if Yes, How?

In Heger and Jeschke (2014), we suggested weighting tests of hypotheses according to the used method (experiments > observation; field studies > enclosures > lab environments) and according to the number of focal species (many focal species > few focal species). Sam Scheiner and Gordon Fox (Chapter 5) argue that this suggestion is arbitrary, and that there are no data supporting the decision that, for example, experimental field studies are given a four times higher weight than observational field studies. In Chapter 4, James Griesemer points out that with our suggestion of assigning different weights to different study approaches, an implicit evaluation becomes an integral part of the data.

We agree that these are issues that need to be discussed. Our basic motivation for suggesting this weighting scheme was that the empirical tests included in the analysis of Heger and Jeschke (2014) differed very much in their approaches, and we felt a need for standardization and for putting weight on particularly relevant tests. Certainly, there are many other possibilities to approach some form of standardization, and the decision if and how to weight studies should depend on the research question. In any case, it will be important to consider the potential influence of the used weighting scheme for the outcome of a meta-analysis *sensu lato* (we use this term in the broad sense of an analysis of data from different empirical studies). Surprisingly, this topic has been ignored by most previous meta-analyses *sensu lato* (although see Norris *et al.*, 2012) and we very much hope our suggestions stimulate discussions about this important question.

We highly appreciate the useful contribution by James Griesemer (Chapter 4) on this topic. He suggests locating the studies testing a hypothesis in 'Levins' space', based on information about each study's approaches and methods. In this way, each study could be assigned a location within the three-dimensional space opened up by axes (1) precision, (2) realism and (3) generality. Levins (1966) was referring to population biological models but it seems reasonable to assume that his idea of a trade-off between these three factors and manageability is applicable to empirical studies as well. An empirical study trying to maximize precision (e.g. by including many replicates), generality (e.g. by considering many focal species) and realism (e.g. by trying to capture field conditions) at the same time would certainly be hard to manage.

We therefore agree that the three axes could well be used to semi-quantitatively characterize empirical studies. In Levins' space, a field study could be assigned a higher value for realism than a study conducted in a laboratory, and a study including many focal species a higher value for generality than a study using a single species. A study with many replicates would have a higher score for precision than a study that has not been replicated. One important component seems to be missing, however, and this is the distinction between pure observation and experimental manipulation. A study conducting accurate and well-replicated surveys in the field taking many species into account is purely observational but could be as precise, general and realistic as a comparable experiment. The three dimensions thus do not seem to be helpful in differentiating between observational and experimental approaches.

On the basis of these considerations, we suggest assigning four separate values to each empirical test in an analysis. These values could represent the level of (1) precision, (2) realism, (3) generality and (4) observation vs experiment. For 1, 2 and 3, these values could, for instance, range from a minimum of 1 to a maximum of 5; for 4, we suggest dichotomous (e.g. 1/4) values (the best range for these values is up for discussion). These values could then be used to produce four different scores for each study instead of one joint score. In subsequent analyses, it could be tested whether the overall results of the meta-analysis change if the studies are weighted according to precision, realism, generality or general approach. This procedure could help, as James Griesemer pointed out (Chapter 4), to test the robustness of the meta-analytic results and to prevent certain biases in study design staying undiscovered, and wrong conclusions being drawn from a meta-analysis. A first test of such an approach will be presented in Chapter 11.

Hierarchy or Network? Hypotheses or Causalities?

In their comment to our approach, Sam Scheiner and Gordon Fox (Chapter 5) point out that hypotheses (or theories and models in their terminology) are not necessarily structured in a hierarchical way. Hypotheses may overlap or may be linked in a network, and theories could be structured in a modular way. We agree with this comment. We do not believe that the hierarchical structure is inherent to ecological theory and that all hypotheses necessarily need to be part of a hierarchy. But we believe that to think of hypotheses (or theories or models) in terms of a hierarchy can be very helpful. We regard it as a methodological approach, not as an inherent structure of theory.

We agree with Scheiner and Fox that a network can also be a very helpful methodological approach (see Chapter 7, this volume). As they correctly point out, invasions as well as other phenomena in ecology are usually not driven by single, but by multiple, oftentimes interacting processes, and usually no single mechanism offers a general explanation for an invasion event (Heger and Trepl, 2003; Catford et al., 2009). We admit that the HoH approach tends to suggest a reductionistic view on invasion processes, and networks in many cases are better able to convey the complex nature of a process or phenomenon. We suggest, however, viewing these two approaches as being complementary, not as mutually exclusive. The hierarchical as well as the network approach are two ways of addressing complexity, and both can be used as a starting point to structure ideas, results, models or hypotheses. With the hierarchical approach, there is a danger of underestimating interactions between processes that are addressed separately, as pointed out by Scheiner and Fox (Chapter 5). We believe that with a careful interpretation of results it is possible to reveal such interactions.

In fact, we believe it can be especially useful to combine these two approaches, thus building hierarchically structured networks. In a hierarchical network of hypotheses, a network of overarching hypotheses could form the top layer, which could be linked to lower layers of working hypotheses and operational hypotheses, and within these lower layers, hypotheses could be linked to each other as well. The goal to create an online portal with a hierarchical network of invasion hypotheses has already been mentioned in Jeschke (2014) (see Chapters 17 and 18, this volume).

A somehow related comment by Scheiner and Fox (Chapter 5) is that it may be more useful to think about causal networks instead of hierarchies of hypotheses. Again, we think these approaches are not mutually exclusive. In the case of invasion ecology, we expect that the HoH approach can be used to improve theory by organizing hypotheses and uncovering complexity – as is done in this book. In a subsequent step, the results of assessments of evidence based on a hierarchy of hypotheses could be fed into a causal network such as the one proposed by Gurevitch et al. (2011; see Fig. 5.1, Chapter 5, this volume), thus continuously improving its accurateness. Also, this or a similar network could be expanded in the third dimension to include refinements of the suggested components, thus forming a hierarchical network.

What Conclusions Can be Drawn from a 'Red' or a 'Green' Branch?

In Jeschke et al. (2012) and Heger and Jeschke (2014), we suggested to combine the HoH approach with an assessment of the usefulness of hypotheses. We would like to point out that the HoH approach does not per se assess the usefulness of hypotheses. It allows to hierarchically structure hypotheses and can also be used to assign empirical studies to different sub-hypotheses (i.e. working or operational hypotheses). The subsequent step of hypotheses testing is one possible modular extension of the core HoH approach but this is optional and not mandatory (see below).

In Jeschke et al. (2012) and Heger and Jeschke (2014), we combined the core HoH approach with this optional extension and suggested to determine the usefulness of a hypothesis (no matter on which level of the hierarchy) based on how much empirical evidence supports or questions it. In a semiquantitative manner, we furthermore suggested a threshold of 50%: if more than half

of the available empirical tests support a (sub-)hypothesis, it is useful and, in the graphical display of results in Heger and Jeschke (2014), we assigned the colour green to such hypotheses. If more than half of the available evidence rejects a hypothesis, we marked it as red, suggesting it is not very useful (see also figures in Chapters 8–10 and 13–16, this volume).

These suggestions have many implications and are debatable. We pointed this out already (see Heger and Jeschke, 2014) and as expected we have received critical comments by colleagues, which in turn oftentimes contradict each other, thus highlighting the need for a wider discussion of this question. The main question here is which consequences negative or positive evidence should have – a question located in the realm of science philosophy. Should the hypotheses, or branches of the hierarchy, that are coloured in red, be discarded, because they are not useful? And are the green ones suitable to be integrated into theory, with no more empirical testing necessary? Should future research particularly focus on those parts of the hierarchy where there is mixed evidence or no clear results available (i.e. the white branches of the hierarchy)? One could also argue that a hypothesis has proven useful if a single study has found evidence supporting it (see Scheiner and Fox, Chapter 5).

According to Karl Popper (1959), though, verification does not work. For instance, it will never be possible to verify the hypothesis that 'all swans are white'. Even if all conducted studies only find white swans, there might still be a black one somewhere. This could mean that no hypothesis should ever be assigned a green colour, and as soon as one study has been found to contradict a hypothesis, this hypothesis should be coloured red and discarded (see also Farji-Brener and Amador-Vargas, Chapter 3, Fig. 3.1). We do not agree that this procedure is adequate for ecology and in our view this is not the way ecological science is conducted today. Also, in science philosophy, this view is not the only one. An approach that might be more useful to reflect ecological science is the one proposed by Imre Lakatos (1970).

According to him, researchers should beware of naïve falsificationism and should not abandon a hypothesis as soon as negative evidence has been found. Instead, ad-hoc hypotheses should be allowed to modify the previous hypothesis, and to incorporate the gained evidence into an evolving body of theory. In line with Thomas Kuhn (1970), he was convinced that science is always guided by scientific paradigms. These paradigms influence which questions are being asked and how results are being interpreted. Lakatos (1970) suggested that a current paradigm should only be abandoned if it can be replaced by a better one.

We agree that some poorly supported invasion hypotheses can probably be rescued by revising them and that there may be no need to abandon them. At the same time, it is important to identify and get rid of zombie ideas (Fox, 2011). There is now a plethora of invasion hypotheses – some of them are similar or overlapping, whereas others contradict each other (e.g. Catford et al., 2009, and Chapter 7, this volume) – and it is highly questionable whether keeping a mess of poorly supported hypotheses is beneficial for the field.

Regarding the HoH approach as applied in Heger and Jeschke (2014), we therefore suggest that flagging hypotheses as more or less useful based on available empirical evidence is a valid approach, indicating which hypotheses should be considered to be revised, restricted in their use (e.g. to certain formulations or taxonomic groups) or abandoned. The red and green coloration (e.g. in Fig. 9.1 in Chapter 9, this volume) can deliver hints on which specific formulations of a broad hypothesis are more promising than others. This view is in line with the suggestion that ecological research should intend to search for rules in the sense of empirical generalizations rather than for laws in a strict sense (Murray, 2004).

An alternative approach to using a classification into more or less useful hypotheses based on a quantification of the level of available evidence is to analyse the robustness of hypotheses as suggested by James Griesemer (Chapter 4, this volume; see also Levins, 1966). To this end, the level of evidence for each hypothesis could be assessed taking into account different scores for the level of (1) precision, (2) realism, (3) generality and (4) observation vs experiment, as described above. It could therefore be assessed whether positive evidence for a hypothesis is only found if a certain research approach is used or a certain study system is considered, or whether positive evidence can be gained across research methods and study systems.

Is the Suggested Approach Vote Counting?

As mentioned above, the HoH approach does not have to, but can be combined with meta-analysis *sensu lato*. For this purpose, the researcher can choose to apply: (i) a semi-quantitative method, e.g. with three ordinal levels (supporting, undecided, questioning) or a different number of levels (e.g. five levels); or (ii) a fully quantitative method based on effect sizes (i.e. meta-analysis *sensu stricto*). Thus far, we have only provided examples for (i) in published applications of the HoH approach (Jeschke et al., 2012; Heger and Jeschke, 2014), which may have led to the wrong impression that it is inherent to the HoH approach. Instead, the HoH approach is a toolbox that helps structuring theory, research and evidence, and that also includes optional tools for meta-analysis *sensu lato*.

Another misconception is that the semi-quantitative method we have used is vote counting, as suggested by Scheiner and Fox in Chapter 5. Vote counting is an approach where the statistical significance of available studies is used to classify studies as supporting or not supporting (Borenstein et al., 2009; Koricheva et al., 2013; Gurevitch and Nakagawa, 2015). Two or three categories are typically used here, and an important drawback of vote counting is that it is based on whether or not the results of available studies were statistically significant. P values should be interpreted carefully (e.g. Nester, 1996; Stephens et al., 2007; and references therein); vote counting based on p values thus has key weaknesses, as was, for example, highlighted by Koricheva *et al.* (2013, p. 6): 'Because vote counting is based on the statistical significance of the research findings, it has low power for effects of relatively small magnitude [. . .]; this is due to the statistical power of small studies possibly being too low to detect an effect.' Null hypothesis significance tests have no symmetrical design. They are instead primarily designed for falsification: the null hypothesis is typically rejected if $p < 0.05$ but it cannot be accepted if $p > 0.05$ (Borenstein *et al.*, 2009). The asymmetric design of significance tests is an important reason why it is not advisable to apply vote counting based on p values. The semi-quantitative approach we have applied does not simply classify studies on the basis of their statistical significance, hence it is not vote counting. It is rather an approach that takes all available evidence into account to classify studies as supporting, undecided or questioning. A primary role in this classification is played by effect sizes, but they are – in contrast to meta-analysis *sensu stricto* – not directly used in the further analysis; instead, they are translated into an ordinal, semi-quantitative score where other relevant information, if any, is considered as well. It is this semi-quantitative score that is used for the further analysis.

Directly using effect sizes in the further analysis and thus applying a fully quantitative meta-analytic approach is preferable in those cases where the results of studies addressing the focal hypothesis can be reasonably expressed in the same effect size metric (including after transformation, e.g. Borenstein *et al.*, 2009). This is not always the case because for broad hypotheses there is often a plethora of different research studies and it might be impossible to express their results in the same effect size metric. Even if it is mathematically possible to express results in the same metric, it is not always clear if this is advisable because effect sizes can sometimes be expected to genuinely vary among sub-hypotheses, taxonomic groups, ecological level and spatiotemporal scales. Thus, the decision of whether to apply a semi-quantitative or fully quantitative approach in assessing empirical support for hypotheses should be done on a case-by-case basis.

The Extended HoH Approach as a Modular Toolbox

In responding to the different comments, we already indicated that we think of the HoH approach as a stimulus and methodological suggestion rather than a fixed method. We use this final paragraph to picture it as a modular toolbox with many possible applications.

The HoH approach can be applied in a purely evidence-driven, as well as in a purely theory-driven, way and also in many other ways that lie in between (Fig. 6.1; see also Griesemer, Chapter 4). The core of this method is the idea that complexity can be approached by structuring the respective topic in a hierarchical way. This core method cannot only be used to structure hypotheses, but also predictions, models, concepts, ideas or research questions. For example, a given overarching research question can be divided into sub-questions; this would then be a hierarchy of questions (i.e. an HoQ; cf. the HoEO in Chapter 3, this volume), following the same logic as described for an HoH.

The methodological core can be nourished by several optional elements (Fig. 6.2). For example, the hierarchy can be supplemented by a network, linking top levels of different hierarchies, or creating links also within lower levels (hierarchical network, see above). Also, the core tool can be combined with a systematic review and meta-analysis *sensu lato* (Fig. 6.2). To prepare a meta-analysis *sensu lato*, the researcher can choose to apply: (i) a semi-quantitative approach, e.g. with three ordinal levels (supporting, undecided, questioning; Heger and Jeschke, 2014) or a different number of levels (e.g. five); or (ii) a fully quantitative approach based on effect sizes (i.e. meta-analysis *sensu stricto*; see the previous section). Similarly, a weighting procedure can be applied when assessing empirical support for hypotheses, e.g. the one proposed in

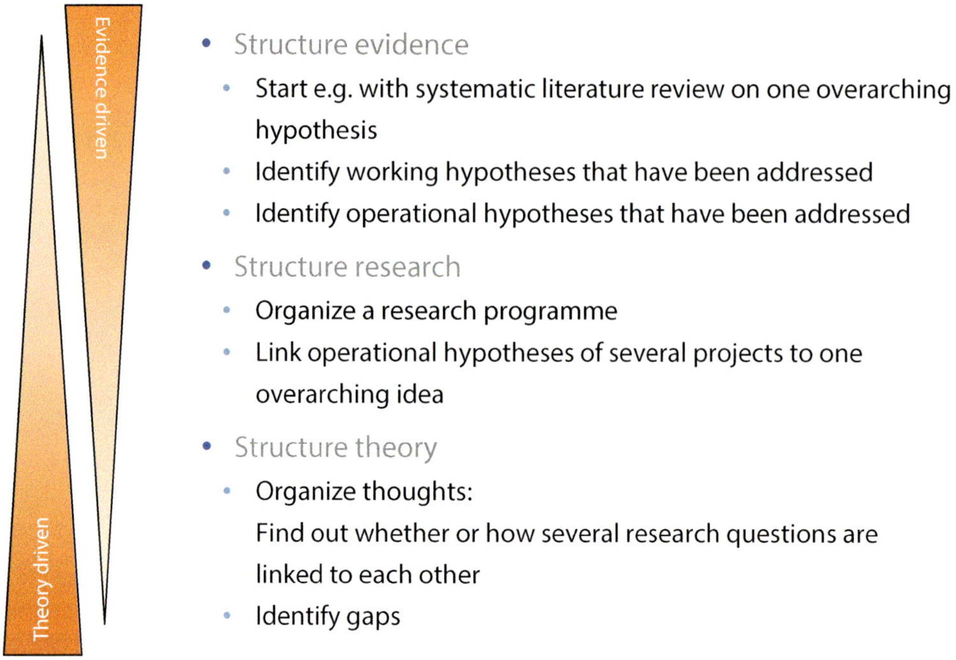

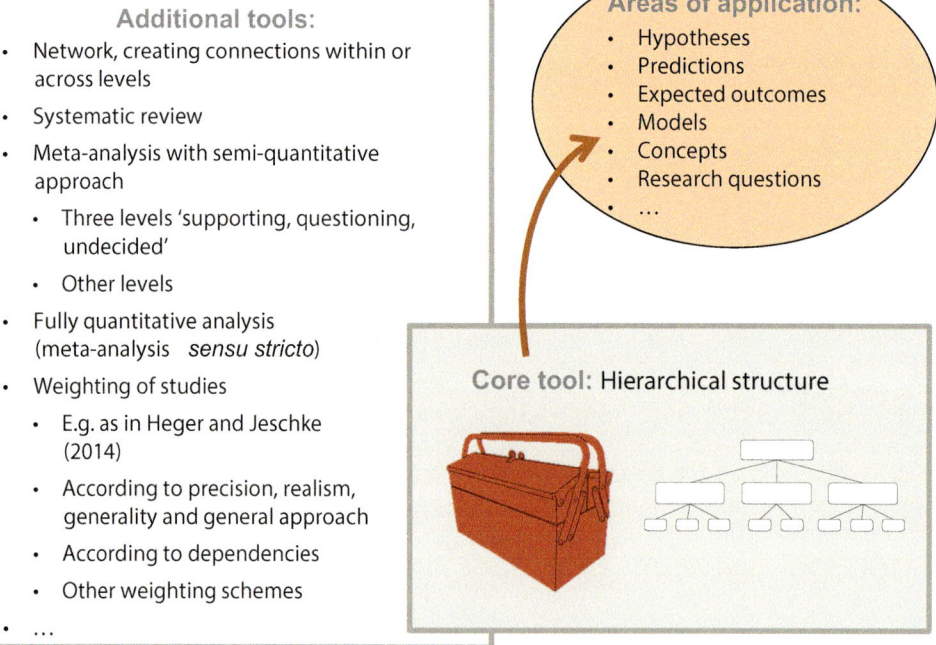

Evidence driven

Theory driven

- Structure evidence
  - Start e.g. with systematic literature review on one overarching hypothesis
  - Identify working hypotheses that have been addressed
  - Identify operational hypotheses that have been addressed

- Structure research
  - Organize a research programme
  - Link operational hypotheses of several projects to one overarching idea

- Structure theory
  - Organize thoughts:
    Find out whether or how several research questions are linked to each other
  - Identify gaps

Fig. 6.1. Some possible applications of the hierarchy-of-hypotheses approach, ranging from purely theory driven to purely evidence driven implementations.

Additional tools:
- Network, creating connections within or across levels
- Systematic review
- Meta-analysis with semi-quantitative approach
  - Three levels 'supporting, questioning, undecided'
  - Other levels
- Fully quantitative analysis (meta-analysis *sensu stricto*)
- Weighting of studies
  - E.g. as in Heger and Jeschke (2014)
  - According to precision, realism, generality and general approach
  - According to dependencies
  - Other weighting schemes
- …

Areas of application:
- Hypotheses
- Predictions
- Expected outcomes
- Models
- Concepts
- Research questions
- …

Core tool: Hierarchical structure

Fig. 6.2. The extended HoH approach as a modular toolbox. The core tool can be combined with additional tools in a modular way.

Heger and Jeschke (2014), the one introduced above, or based on a preceding assessment of the reliability or quality of each study using the method proposed in Mupepele *et al.* (2016) (see Norris *et al.*, 2012, for more ideas). The analysis could also correct for dependencies among studies, e.g. due to authorship (Lokatis and Jeschke, 2018) or more obvious financial or political conflicts of interest (e.g. funding of a given study by an organization with certain interests in the study's outcome).

On the basis of the comments received in Chapters 3–5, we here presented some new ideas on how to apply the HoH approach and we clarified our view of the approach as a modular toolbox. The core tool helps structuring theory, research and evidence, particularly by dividing an overarching hypothesis (or research question or something similar) into working and operational (sub-)hypotheses. Which additional modules or tools are applied, and in which way they are implemented, very much depends on the purpose of a given study. In the following chapters, we will apply some combinations of the suggested modules, hoping to stimulate creative and versatile applications of the HoH approach in the future.

References

Borenstein, M., Hedges, L.V., Higgins, J.P.T. and Rothstein, H.R. (2009) *Introduction to Meta-analysis.* Wiley, Chichester, UK.

Catford, J.A., Jansson, R. and Nilsson, C. (2009) Reducing redundancy in invasion ecology by integrating hypotheses into a single theoretical framework. *Diversity and Distributions* 15, 22–40.

Fox, J. (2011) Zombie ideas in ecology. *Oikos blog.* Available at: http://oikosjournal.wordpress.com/2011/06/17/zombie-ideas-in-ecology/ (accessed 29 September 2017).

Giere, R. (1988) *Explaining Science: a Cognitive Approach.* University of Chicago Press, Chicago, Illinois.

Giere, R. (1997) *Understanding Scientific Reasoning,* 4th edn. Harcourt, Brace, Jovanovich, Fort Worth, Texas.

Gurevitch, J. and Nakagawa, S. (2015) Research synthesis methods in ecology. In: Fox, G.A., Negrete-Yankelevich, S. and Sosa, V.J. (eds) *Ecological Statistics: Contemporary Theory and Application.* Oxford University Press, Oxford, UK, pp. 200–227.

Gurevitch, J., Fox, G. A., Wardle, G. M., Inderjit and Taub, D. (2011) Emergent insights from the synthesis of conceptual frameworks for biological invasions. *Ecology Letters* 14, 407–418.

Heger, T. and Trepl, L. (2003) Predicting biological invasions. *Biological Invasions* 5, 313–321.

Heger, T. and Jeschke, J.M. (2014) The enemy release hypothesis as a hierarchy of hypotheses. *Oikos* 123, 741–750.

Jeschke, J.M. (2014) General hypotheses in invasion ecology. *Diversity and Distributions* 20, 1229–1234.

Jeschke, J.M., Gómez Aparicio, L., Haider, S., Heger, T., Lortie, C. J., Pyšek, P. and Strayer, D.L. (2012) Support for major hypotheses in invasion biology is uneven and declining. *NeoBiota* 14, 1–20.

Koricheva, J., Gurevitch, J. and Mengerson, K. (eds) (2013) *Handbook of Meta-analysis in Ecology and Evolution.* Princeton University Press, Princeton, New Jersey.

Kuhn, T.S. (1970) *The Structure of Scientific Revolutions.* 2nd edn. University of Chicago Press, Chicago, Illinois.

Lakatos, I. (1970) *Criticism and the Growth of Knowledge.* Cambridge University Press, New York.

Levins, R. (1966) The strategy of model building in population biology. *American Scientist* 54, 421–431.

Lokatis, S. and Jeschke, J.M. (2018) The island rule: An assessment of biases and research trends. *Journal of Biogeography* 45, 289–303.

Mupepele, A.-C., Walsh, J.C., Sutherland, W.J. and Dormann, C.F. (2016) An evidence assessment tool for ecosystem services and conservation studies. *Ecological Applications* 26, 1295–1301.

Murray, B.G.J. (2004) Laws, hypotheses, guesses. *The American Biologist Teacher* 66, 598–599.

Nester, M.R. (1996) An applied statistician's creed. *Applied Statistics*, 45, 401–410.

Norris, R.H., Webb, J.A., Nichols, S.J., Stewardson, M.J. and Harrison, E.T. (2012) Analyzing cause and effect in environmental assessments: using weighted evidence from the literature. *Freshwater Science* 31, 5–21.

Popper K. (1959) *The Logic of Scientific Discovery.* Basic Books, New York.

Stephens, P.A., Buskirk, S.W. and Martínez del Rio, C. (2007) Inference in ecology and evolution. *Trends in Ecology & Evolution* 22, 192–197.

Part II

Hypothesis Network and 12 Focal Hypotheses

7

A Network of Invasion Hypotheses

Martin Enders[1,2,3]* and Jonathan M. Jeschke[1,2,3]

[1]Freie Universität Berlin, Institute of Biology, Berlin, Germany; [2]Leibniz-Institute of Freshwater Ecology and Inland Fisheries (IGB), Berlin, Germany; [3]Berlin-Brandenburg Institute of Advanced Biodiversity Research (BBIB), Berlin, Germany

Abstract

Hypotheses of research disciplines are typically not isolated from each other but share similarities. In a broad sense as defined here, they form an important part of the theoretical–conceptual understanding of a given topic, e.g. invasion hypotheses *sensu lato* represent an important part of our understanding of biological invasions. Dynamic research disciplines such as invasion biology have so many hypotheses that it is even hard for experts to keep track, and researchers from other disciplines as well as policymakers, managers and other interested people find it extremely complicated to get to grips with invasion hypotheses. To tackle this situation, we argue that it is useful to define key hypotheses and visualize their relationships. We define 35 of the arguably most common invasion hypotheses and outline three approaches to create hypothesis networks that visualize the similarities and dissimilarities between hypotheses: (i) the bibliometric approach; (ii) the survey approach; and (iii) the matrix approach. The latter approach is in the focus of this chapter. It is centred around a matrix that represents the characteristics or traits of each hypothesis. Here we assigned such traits to 35 invasion hypotheses based on 13 trait categories. We then calculated the similarities between them and created a hypothesis network visualizing these similarities. With the same trait matrix, we created a smaller

network focused on the 12 hypotheses featured in this book. This network thus illustrates the relationships between these 12 hypotheses and can be used as a map for the following chapters.

Introduction

It is said that before Napoleon Bonaparte went into battle he sat in a big sandbox, planning all his battle moves with miniature figures in advance (Botham, 2006). Sure, Napoleon took it too far, but the concept of visualizing a battle on a map of the surroundings wasn't that extravagant. This is also the idea behind networks of invasion biology: to see the bigger picture behind it, the connections, similarities and dissimilarities at once, to plan your next move – in this case regarding research or management of biological invasions. To see the bigger picture in the field of invasion biology is getting more and more important, especially when considering the progress of the field in the last 25 years. We have reached a point at which we produce more information every day but seem to have lost the general overview of the field. This is why the field of invasion biology needs something to order it, for example a map of the field. Networks seem to be promising tools to create useful maps of this and other research fields (Jeschke, 2014).

* Corresponding author. E-mail: enders.martin@gmx.net

There are different possible approaches to create networks, each of them with their own benefits and disadvantages. We outline three such approaches (this list is, of course, not exhaustive) and will concentrate on the third one of these for the remainder of this chapter:

1. *Bibliometric approach*: Here, the full text or meta-data of publications are analysed and used to build a network based on citations, co-citations or collaborations between working groups, or on content similarity by comparing key phrases. The nodes in such a network can be authors, journals or (sub-) disciplines. See de Solla Price (1965) or Börner (2010) for examples of applications of this approach.

2. *Survey approach*: The idea here is to use the judgement of different experts on the similarities and dissimilarities of hypotheses in a research field. This procedure was already used for the field of invasion biology by Enders *et al.* (2018) who developed an online questionnaire asking experts in the field about similarities and dissimilarities between 33 common invasion hypotheses. Each hypothesis was defined and a short explanation provided (based on and extending Catford *et al.*, 2009; a further extension is provided in this chapter, see Table 7.1). Participants were asked to choose up to three hypotheses that they know best. The survey then took the chosen hypotheses and randomly paired them with other hypotheses; the participants were asked about their similarities or dissimilarities. From the results, different networks were created using different formulae for calculating between-hypotheses similarity.

3. *Matrix approach*: This approach compares characteristics (i.e. traits) of ideas in a specific field. If two hypotheses share a number of characteristics beyond a certain threshold, they are termed 'similar', and a connection between these hypotheses is drawn in the network. This approach will be explained in detail in the next section using invasion hypotheses as an example.

All approaches can be applied in any field of research where a number of key hypotheses exist. It is quite surprising that such approaches have only been rarely applied, even though Naisbitt's quote 'we are drowning in information but starved for knowledge' is over three decades old now (p. 24 in Naisbitt, 1982). We urgently need tools to synthesize the increasing amounts of information in order to make them more accessible and usable. It is the goal of this chapter to contribute to the development of such tools by focusing on one possible approach, the one we termed 'matrix approach'.

Methods

Applying the matrix approach, we first defined 35 common invasion hypotheses. This list of definitions was based on Catford *et al.* (2009), references cited therein and further sources cited in Table 7.1. For Enders *et al.* (2018) we extended the list provided in Catford *et al.* (2009); for this chapter we extended it once more, resulting in the list provided in Table 7.1. Please note, however, that there are further invasion hypotheses (see e.g. Chapter 17, this volume, and Ricciardi *et al.*, 2013).

We then developed a matrix containing traits for each hypothesis in 13 categories (Table 7.2). The category 'lag time' describes the time period that the mechanism or effect represented by a hypothesis needs, starting from the introduction of a non-native species. We differentiated very short (++), relatively short (+), intermediate (+−), relatively long (−) and very long (− −) lag times. The next two categories were 'propagule pressure' and 'other human actions' and are summarized under human interference. These and all following categories were classified as either very important (++), somewhat important (+) or not important (empty cell) for a given hypothesis. The following three categories describe ecosystem properties: 'habitat modification', caused by either humans, non-native or native species or abiotic factors; available 'resources' in the new ecosystem; and other 'ecosystem properties'. The following three categories come under the header biotic interactions: 'enemies', 'mutualism' and 'competition'. They

Table 7.1. List of 35 common invasion hypotheses and how we defined them.

Hypothesis	Definition	Key reference(s)
Adaptation (ADP)	The invasion success of non-native species depends on the adaptation to the conditions in the exotic range before and/or after the introduction. Non-native species that are related to native species are more successful in this adaptation.	Duncan and Williams (2002)
Biotic acceptance aka 'the rich get richer' (BA)	Ecosystems tend to accommodate the establishment and coexistence of non-native species despite the presence and abundance of native species.	Stohlgren et al. (2006)
Biotic indirect effects (BID)	Non-native species benefit from different indirect effects triggered by native species.	Callaway et al. (2004)
Biotic resistance aka diversity-invasibility hypothesis (BR)	An ecosystem with high biodiversity is more resistant against non-native species than an ecosystem with lower biodiversity.	Elton (1958), Levine and D'Antonio (1999)
Darwin's naturalization (DN)	The invasion success of non-native species is higher in areas that are poor in closely related species than in areas that are rich in closely related species.	Darwin (1859)
Disturbance (DS)	The invasion success of non-native species is higher in highly disturbed than in relatively undisturbed ecosystems.	Elton (1958), Hobbs and Huenneke (1992)
Dynamic equilibrium model (DEM)	The establishment of a non-native species depends on natural fluctuations of the ecosystem, which influences the competition of local species.	Hutson (1979)
Empty niche (EN)	The invasion success of non-native species increases with the availability of empty niches in the exotic range.	MacArthur (1970)
Enemy inversion (EI)	Introduced enemies of non-native species are less harmful for them in the exotic than the native range, owing to altered biotic and abiotic conditions.	Colautti et al. (2004)
Enemy of my enemy aka accumulation-of-local-pathogens hypothesis (EE)	Introduced enemies of a non-native species are less harmful to the non-native as compared to the native species.	Eppinga et al. (2006)
Enemy reduction (ERD)	The partial release of enemies in the exotic range is a cause of invasion success.	Colautti et al. (2004)
Enemy release (ER)	The absence of enemies in the exotic range is a cause of invasion success.	Keane and Crawley (2002)
Environmental heterogeneity (EVH)	The invasion success of non-native species is high if the exotic range has a highly heterogeneous environment.	Melbourne et al. (2007)
Evolution of increased competitive ability (EICA)	After having been released from natural enemies, non-native species will allocate more energy in growth and/or reproduction (this re-allocation is due to genetic changes), which makes them more competitive.	Blossey and Nötzold (1995)
Global competition (GC)	A large number of different non-native species is more successful than a small number.	Colautti et al. (2006)

continued

Table 7.1. *continued*

Hypothesis	Definition	Key reference(s)
Habitat filtering (HF)	The invasion success of non-native species in the new area is high if they are pre-adapted to this area.	Darwin (1859)
Human commensalism (HC)	Species that are living in close proximity to humans are more successful in invading new areas than other species.	Jeschke and Strayer (2006)
Ideal weed (IW)	The invasion success of a non-native species depends on its specific traits (e.g. life-history traits).	Elton (1958), Reimánek and Richardson (1996)
Increased resource availability (IRA)	The invasion success of non-native species increases with the availability of resources.	Sher and Hyatt (1999)
Increased susceptibility (IS)	If a non-native species has a lower genetic diversity than the native species, there will be a low probability that the non-native species establishes itself.	Colautti *et al.* (2004)
Invasional meltdown (IM)	The presence of non-native species in an ecosystem facilitates invasion by additional species, increasing their likelihood of survival or ecological impact.	Simberloff and Von Holle (1999), Sax *et al.* (2007)
Island susceptibility hypothesis (ISH)	Non-native species are more likely to become established and have major ecological impacts on islands than on continents.	Jeschke (2008)
Limiting similarity (LS)	The invasion success of non-native species is high if they strongly differ from native species and it is low if they are similar to native species.	MacArthur and Levins (1967)
Missed mutualisms (MM)	In their exotic range, non-native species suffer from missing mutualists.	Mitchell *et al.* (2006)
New associations (NAS)	New relationships between non-native and native species can positively or negatively influence the establishment of the non-native species.	Colautti *et al.* (2004)
Novel weapons (NW)	In the exotic range, non-native species can have a competitive advantage against native species because they possess a novel weapon, i.e. a trait that is new to the resident community of native species and therefore affects them negatively.	Callaway and Ridenour (2004)

Hypothesis	Description	Reference
Opportunity windows (OW)	The invasion success of non-native species increases with the availability of empty niches in the exotic range and the availability of these niches fluctuates spatio-temporally.	Johnstone (1986)
Plasticity hypothesis (PH)	Invasive species are more phenotypically plastic than non-invasive or native ones.	Richards *et al.* (2006)
Propagule pressure (PP)	A high propagule pressure (a composite measure consisting of the number of individuals introduced per introduction event and the frequency of introduction events) is a cause of invasion success.	Lonsdale (1999), Lockwood *et al.* (2013)
Reckless invader aka 'boom-bust' (RI)	A non-native species that is highly successful shortly after its introduction can get reduced in its population or even extinct over time due to different reasons (such as competition with other introduced species or adaptation by native species).	Simberloff and Gibbons (2004)
Resource-enemy release (RER)	The non-native species is released from its natural enemies and can spend more energy in its reproduction, and invasion success increases with the availability of resources.	Blumenthal (2006)
Sampling (SP)	A large number of different non-native species is more likely to become invasive than a small number owing to interspecific competition. Also the species identity of the locals is more important than the richness in terms of the invasion of an area.	Crawley *et al.* (1999)
Shifting defence hypothesis (SDH)	After having been released from natural specialist enemies, non-native species will allocate more energy in cheap (energy-inexpensive) defences against generalist enemies and less energy in expensive defences against specialist enemies (this re-allocation is due to genetic changes); the energy gained in this way will be invested in growth and/or reproduction, which makes the non-native species more competitive.	Doorduin and Vrieling (2011)
Specialist–generalist (SG)	Non-native species are more successful in a new region if the local predators are specialists and local mutualists are generalists.	Callaway *et al.* (2004)
Tens rule (TEN)	Approximately 10% of species successfully take consecutive steps of the invasion process.	Williamson and Brown (1986), Williamson (1996)

Table 7.2. Matrix with traits of 35 invasion hypotheses in 13 different categories.

Hypothesis	1	2	3	4	5	6	7	8	9	10	11	12	13
		Human interference		Ecosystem properties			Biotic interactions			Invader traits			
	Lag Time	Propagule pressure	Other human actions	Habitat modification	Resources	Other ecosystem properties	Enemies	Mutualism	Competition	Phylogenetic distance	Functional novelty	Evolution	Other invader traits
Adaptation [ADP]	+	+		+	+		+	+	+	++	++		++
Biotic acceptance [BA]	+-			++	++	++	+	+	+	++	++		
Biotic indirect effects [BID]	--	+		+	+	++	++	++	++				
Biotic resistance [BR]	++						++		++				
Dynamic equilibrium model [DEM]	+-	+	+	+	+	++			++				+
Darwin's Naturalization [DN]	+-		+	++	+	++	++	++	++	++	++	+	
Disturbance [DS]	+-	+	++	+	++	++							
Enemy of my enemy [EE]	-	+					++		++				
Enemy Inversion [EI]	++		+	+	+	+	+	+	+	+	+	++	
Evolution of increased competitive ability [EICA]	-						++		++	+	+	++	++
Empty niche [EN]	++	+	+	++	+	++	++		++	+	+		++
Enemy release [ER]	++	+		+	+	+	++		+	+	+		
Enemy reduction [ERD]	++						++						
Environmental heterogeneity [EVH]	+-	+		+	+	+	+	+	+	+			++
Global competition [GC]	+-	++	++	++	++	++		+	++	+	+		
Human commensalism [HC]	+-	++	++	+	++	++							
Habitat filtering [HF]	++	+		+			+	+	+	++	++	++	++
Invasional meltdown [IM]	--	++	+	++	++	++	+	++	++				
Increased resource availability [IRA]	++	+	+	++	++	++			++				++
Increased susceptibility [IS]	++	+					+		+				
Island susceptibility hypothesis [ISH]	+-		+		+	+	++	+	+	++	++	+	++
Ideal weed [IW]	++	+						+		+	+		++
Limiting similarity [LS]	+-	+						+	+	++	++	+	+
Missed mutualisms [MM]	+-	+						++					
New associations [NAS]	+-	+					+	+	+				
Novel weapons [NW]	++	+					++		++	++	++	++	++
Opportunity windows [OW]	++		+	+	++	++							++
Plasticity hypothesis [PH]	++						++	++	++			++	++
Propagule pressure [PP]	+-	++			++	++	++		++				
Resource-enemy release [RER]	-				+	+	+	+	+		+	+	+
Reckless invader [RI]	+-				++	++	++	++	++	++	++	+	+
Shifting defence hypothesis [SDH]	+	+					+						
Specialist-generalist [SG]	+-	++						++	++				
Sampling [SP]	+-	++						+	++				
Tens rule [TEN]	+-	++											

describe to which degree each type of inter-action is important for the mechanism or effect represented by a hypothesis. Finally, four categories represent different invader traits: 'phylogenetic distance' between the non-native and resident species; 'functional novelty' of the non-native species, e.g. based on the concept of eco-evolutionary experi-ence (Saul et al., 2013; Saul and Jeschke, 2015); 'evolution' of the non-native species after its introduction; and 'other invader traits' (e.g. life-history traits).

We assessed the similarity of each hypothesis with every other hypothesis by calculating the percentage of shared traits. For this calculation, we excluded categories where both hypotheses had empty cells because an empty cell for both hypotheses cannot be considered a shared trait. If a threshold was reached, a connection was made. In this way, we created two networks. The first network includes all 35 evaluated hypotheses, and connections were made if two hypotheses shared at least 25% of their traits. We clustered the network using yEd's (2017) natural groups algorithm (which is based on the edge-betweenness clustering method proposed by Girvan and Newman, 2005). This resulted in four clusters, three of which were large and one (containing propa-gule pressure, global competition, missed mutualisms and the tens rule) small. For clarity, we included the one small cluster in the closest bigger cluster. The second net-work focused on the 12 hypotheses featured in the following chapters of this book – here, thick connections were made for a threshold of 25% and thin connections for a threshold of 15%.

Results and Discussion

Matrix network for 35 invasion hypotheses

The matrix network (Fig. 7.1) contains a total of 35 nodes (= number of hypotheses) and 151 edges (connections) between them. The average number of connections for a hypothesis is 8.6 ± 2.46 (SD). The three most connected hypotheses (with the highest degree centrality) are resource-enemy release (13 connections), environmental heterogeneity and new associations (both with 12 connections). Enemy reduction and enemy release are the hypotheses with the fewest connections (four connections).

All hypotheses in the purple group (see Fig. 7.1) consider human interference to be important or very important ('+' or '++' in columns 2 or 3 in Table 7.2). Taking all hypotheses, most of those hypotheses that consider human interference to be very important ('++' in columns 2 or 3 in Table 7.2) are in the network's purple group: dis-turbance, global competition, human com-mensalism, invasional meltdown, propagule pressure and the tens rule; only the sam-pling hypothesis is not in the purple group, yet it has as many connections to hypothe-ses in the purple group as it has to hypothe-ses in the red group, thus it is between these two groups. All of the hypotheses in the red cluster consider mutualism as an important factor: all 10 hypotheses for which this is the case ('+' in column 8 in Table 7.1) are included in the red group, which does not contain any other hypotheses. The green cluster includes most hypotheses that con-sider enemies (predators or parasites) to be particularly important: 10 of the 13 hypoth-eses for which this is the case ('++' in column 7 in Table 7.1) are in this group. Out of the 11 hypotheses in this group, only one hypothesis does not explicitly consider enemies to be very important. This hypo-thesis – ideal weed – is characterized by very short time lags, which is also true for other hypotheses in the green group. Thus, the green cluster mainly includes hypothe-ses with a focus on enemies, or the lack thereof, of non-native species and on short time lags.

Matrix network for the 12 hypotheses featured in this book

The matrix network of the 12 hypotheses (Fig. 7.2) contains 27 edges. Darwin's natu-ralization hypothesis and the phenotypic

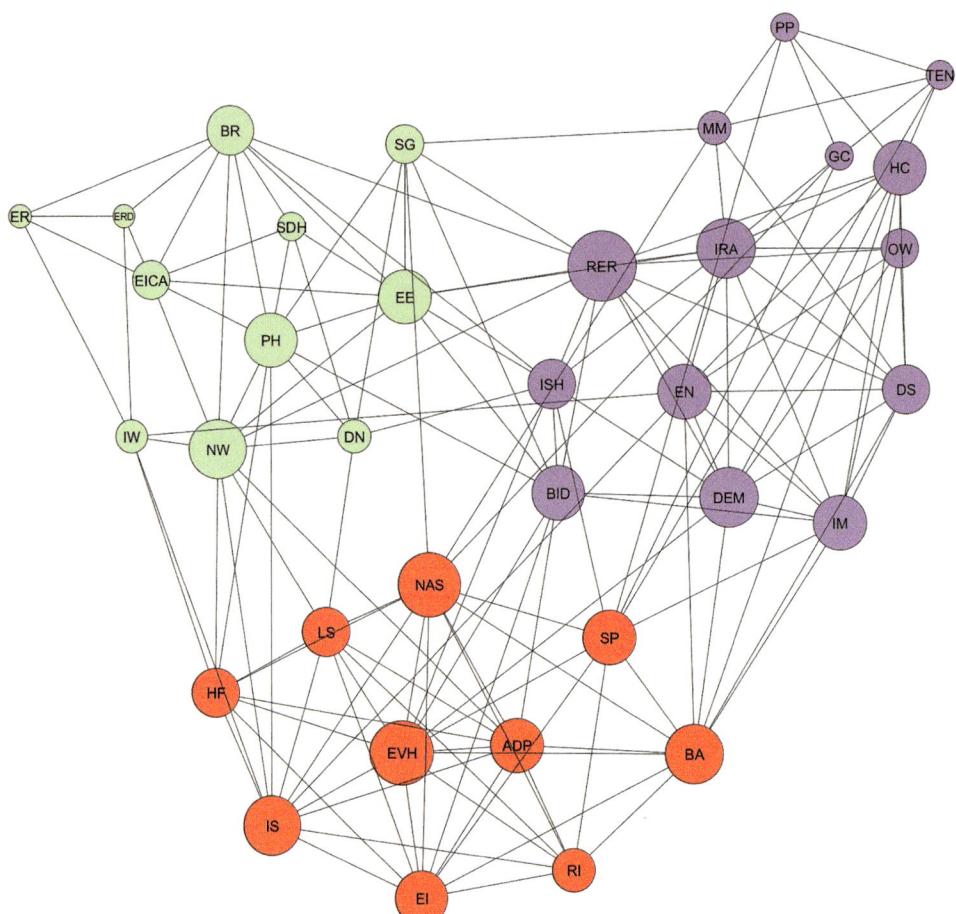

Fig. 7.1 Network with all 35 hypotheses evaluated in this chapter. Connected hypotheses share at least 25% of their traits (the exact spatial position of each hypothesis is arbitrary). Groups of hypotheses are represented by different colours and the size of each circle represents the degree centrality of the hypothesis. Hypothesis names are abbreviated as follows: ADP=adaptation, BA=biotic acceptance, BID=biotic indirect effects, BR=biotic resistance, DN=Darwin's naturalization, DS=disturbance, DEM=dynamic equilibrium, EN=empty niche, EI=enemy inversion, EE=enemy of my enemy, ERD=enemy reduction, ER=enemy release, EVH=environmental heterogeneity, EICA=evolution of increased competitive ability, GC=global competition, HF=habitat filtering, HC=human commensalism, IW=ideal weed, IRA=increased resource availability, IS=increased susceptibility, IM=invasional meltdown, ISH=island susceptibility hypothesis, LS=limiting similarity, MM=missed mutualisms, NAS=new associations, NW=novel weapons, OW=opportunity windows, PH=plasticity hypothesis, PP=propagule pressure, RI=reckless invader, RER=resource-enemy release, SP=sampling, SDH=shifting defence hypothesis, SG=specialist–generalist, TEN=tens rule.

plasticity hypothesis (both seven connections) are the most connected hypotheses (i.e. with the highest degree centrality) in this smaller network. In contrast, limiting similarity, propagule pressure and the tens rule are the least connected hypotheses with two connections each. All of these 12 hypotheses will be explored in detail in the following chapters, and we will come back to this network in the synthesizing Chapter 17.

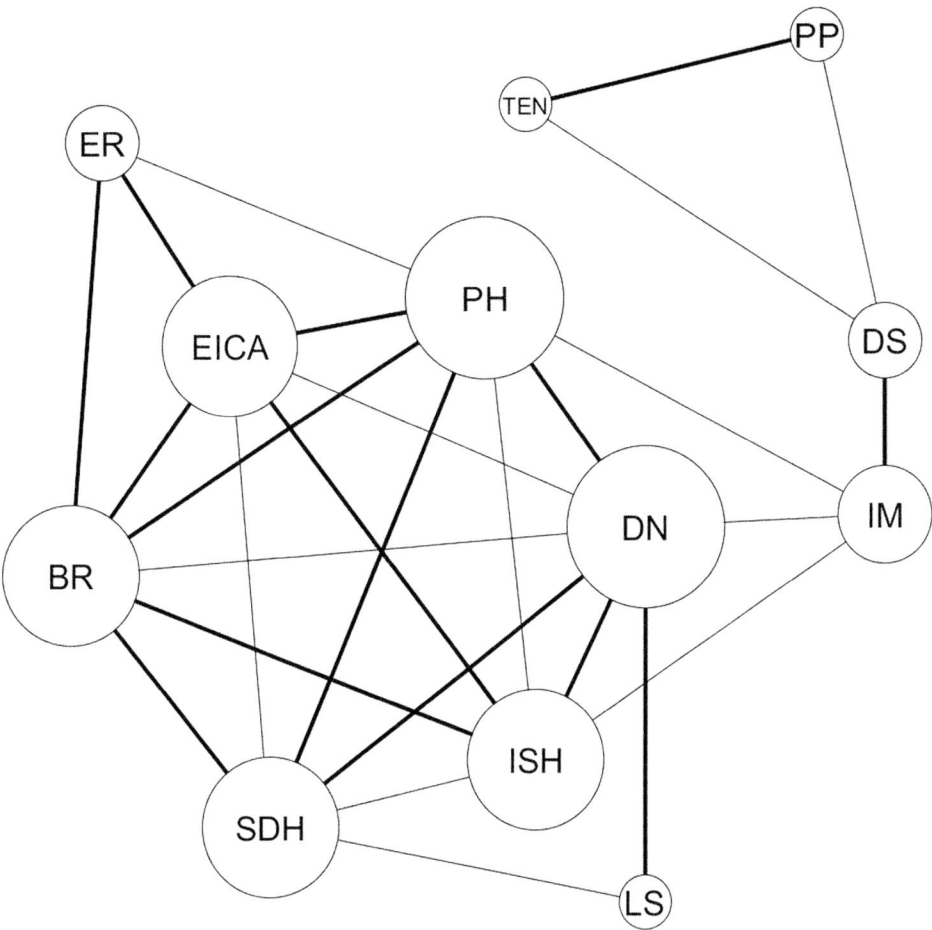

Fig. 7.2. Network with the 12 hypotheses featured in this book. Connected hypotheses share at least 15% of their traits, and connections are thick if they share at least 25% of their traits (otherwise as Fig. 7.1).

Conclusions

The benefits of hypothesis networks are obvious: in addition to providing a map with the main hypotheses of a field, the most central and connected hypotheses can be easily identified. Furthermore, they can convey, depending on the way they are created, much additional valuable information, e.g. hidden similarities among hypotheses, contradictions between hypotheses, thematic groups and, if the release or 'birth date' of hypotheses is considered, developments in the field: when were hypotheses born, which branches of hypothesis networks are particularly thriving and which ones are basically dead ends?

Yet there is need for much additional research on approaches for creating such networks and on the best way to interpret them. For instance, one can see from the results presented in this chapter that it depends on the number of nodes which hypotheses are the ones with the highest degree centrality. Also, there are different methods to create hypothesis networks as mentioned in the Introduction section. Each of these methods can be specified and fitted depending on the specific goals one wants to achieve with the network, e.g. a bibliometric network can be created based on simple co-authorships or co-citations but also based on full publication texts. The same is true for the survey approach: in Enders *et al.* (2018), we used different mathematical metrics to create different networks based on the survey results. Also, the characteristics of the hypotheses given in Table 7.1 can be mathematically analysed in different ways to create yet different hypothesis networks. Thus, one can imagine numerous, apparently reasonable approaches to create hypothesis networks. We have just started to explore this issue and hope others will join us in order to identify the most useful approaches depending on the goal one wants to achieve with a given network. More generally speaking, hypothesis networks seem to be one promising tool for synthesizing the increasing amounts of information in research fields such as invasion biology, yet they need to be further developed and combined with additional synthesis tools.

References

Blumenthal, D.M. (2006) Interactions between resource availability and enemy release in plant invasion. *Ecology Letters* 9, 887–895.

Blossey, B. and Nötzold, R. (1995) Evolution of increased competitive ability in invasive non-indigenous plants: a hypothesis. *Journal of Ecology* 83, 887–889.

Börner K. (2010) *Atlas of Science – Visualizing What we Know.* MIT Press, London.

Botham, N. (2006) *The book of useless information.* Berkeley, London.

Callaway, R.M. and Ridenour, W.M. (2004) Novel weapons: invasive success and the evolution of increased competitive ability. *Frontiers in Ecology and the Environment* 2, 436–443.

Callaway, R.M., Thelen, G.C., Rodriguez, A. and Holben, W.E. (2004) Soil biota and exotic plant invasion. *Nature* 427, 731–733.

Catford, J.A., Jansson, R. and Nilsson, C. (2009) Reducing redundancy in invasion ecology by integrating hypotheses into a single theoretical framework. *Diversity and Distributions* 15, 22–40.

Colautti, R.I., Ricciardi, A., Grigorovich, I.A. and MacIsaac, H.J. (2004) Is invasion success explained by the enemy release hypothesis? *Ecology Letters* 7, 721–733.

Colautti, R., Grigorovich, I. and MacIsaac, H. (2006) Propagule pressure: a null model for biological invasions. *Biological Invasions* 8, 1023–1037.

Crawley, M.J., Brown, S.L., Heard, M.S. and Edwards, G.R. (1999) Invasion-resistance in experimental grassland communities: species richness or species identity? *Ecology Letters* 2, 140–148.

Darwin, C. (1859) *On the Origin of Species by Means of Natural Selection, or the Preservation of Favoured Races in the Struggle for Life.* Murray, London.

de Solla Price, D.J. (1965) Networks of scientific papers. *Science* 149, 510–515.

Doorduin, L.J. and Vrieling, K. (2011) A review of the phytochemical support for the shifting defence hypothesis. *Phytochemistry Reviews* 10, 99–106.

Duncan, R.P. and Williams, P.A. (2002) Ecology: Darwin's naturalization hypothesis challenged. *Nature* 417, 608–609.

Elton, C.S. (1958) *The Ecology of Invasions by Animals and Plants.* Methuen, London.

Enders, M., Hütt, M.-T. and Jeschke, J.M. (2018) Drawing a map of invasion biology based on a network of hypotheses. *Ecosphere.* DOI: 10.1002/ecs2.2146.

Eppinga, M.B., Rietkerk, M., Dekker, S.C., De Ruiter, P.C. and Van der Putten, W.H. (2006) Accumulation of local pathogens: a new hypothesis to explain exotic plant invasions. *Oikos*, 114, 168–176.

Girvan, M. and Newman, M.E.J. (2005) Community structure in social and biological networks. *Proceedings of the National Academy of Sciences USA* 99, 7821–7826.

Hobbs, R.J. and Huenneke, L.F. (1992) Disturbance, diversity, and invasion: implications for conservation. *Conservation Biology* 6, 324–337.

Hutson, M.A. (1979) A general hypothesis of species diversity. *American Naturalist* 113, 81–101.

Jeschke, J.M. (2008) Across islands and continents, mammals are more successful invaders than birds. *Diversity and Distributions* 14, 913–916.

Jeschke, J.M. (2014) General hypotheses in invasion ecology. *Diversity and Distributions*, 20, 1229–1234.

Jeschke, J.M. and Strayer, D.L. (2006) Determinants of vertebrate invasion success in Europe and North America. *Global Change Biology* 12, 1608–1619.

Johnstone, I.M. (1986) Plant invasion windows: a time-based classification of invasion potential. *Biological Reviews* 61, 369–394.

Keane, R.M. and Crawley, M.J. (2002) Exotic plant invasions and the enemy release hypothesis. *Trends in Ecology & Evolution* 17, 164–170.

Levine, J.M. and D'Antonio, C.M. (1999) Elton revisited: A review of evidence linking diversity and invisibility. *Oikos* 87, 15–26.

Lockwood, J.L., Hoopes, M.F. and Marchetti, M.P. (2013) *Invasion ecology*, 2nd edn. Wiley-Blackwell, Chichester, UK.

Lonsdale, W.M. (1999) Global patterns of plant invasions and the concept of invasibility. *Ecology* 80, 1522–1536.

MacArthur, R. and Levins, R. (1967) The limiting similarity, convergence, and divergence of coexisting species. *American Naturalist* 101, 377–385.

MacArthur, R. H. (1970) Species packing and competitive equilibrium for many species. *Theoretical Population Biology* 1, 1–11.

Melbourne, B.A., Cornell, H.V., Davies, K.F., Dugaw, C.J., Elmendorf, S., Freestone, A.L., Hall, R.J., Harrison, S., Hastings, A., Holland, M. *et al.* (2007) Invasion in a heterogeneous world: resistance, coexistence or hostile takeover? *Ecology Letters* 10, 77–94.

Mitchell, C.E., Agrawal, A.A., Bever, J.D., Gilbert, G.S., Hufbauer, R.A., Klironomos, J.N., Maron, J.L., Morris, W.F., Parker, I.M., Power, A.G. *et al.* (2006) Biotic interactions and plant invasions. *Ecology Letters* 9, 726–740.

Naisbitt, J. (1982) *Megatrends: Ten New Directions Transforming our Lives.* Warner Books, New York.

Reimánek, M. and Richardson, D.M. (1996) What attributes make some plant species more invasive? *Ecology* 77, 1655–1661.

Ricciardi, A., Hoopes, M.F., Marchetti, M.P. and Lockwood, J.L. (2013) Progress toward understanding the ecological impacts of nonnative species. *Ecological Monographs* 83, 263–282.

Richards, C.L., Bossdorf, O., Muth, N.Z., Gurevitch, J. and Pigliucci, M. (2006) Jack of all trades, master of some? On the role of phenotypic plasticity in plant invasions. *Ecology Letters* 9, 981–993.

Saul, W.-C., Jeschke, J.M. (2015) Eco-evolutionary experience in novel species interactions. *Ecology Letters* 18, 236–245.

Saul, W.-C., Jeschke, J.M. and Heger, T. (2013) The role of eco-evolutionary experience in invasion success. *NeoBiota* 17, 57–74.

Sax, D.F., Stachowicz, J.J., Brown, J. H., Bruno, J.F., Dawson, M.N., Gaines, S.D., Grosberg, R.K., Hastings, A., Holt, R.D., Mayfield, M. M. *et al.* (2007) Ecological and evolutionary insights from species invasions. *Trends in Ecology & Evolution* 22, 465–471.

Sher, A.A. and Hyatt, L.A. (1999) The disturbed resource-flux invasion matrix: a new framework for patterns of plant invasion, *Biological Invasions* 1, 107–114.

Simberloff, D. and Gibbons, L. (2004) Now you see them, now you don't! – Population crashes of established introduced species. *Biological Invasions* 6, 161–172.

Simberloff, D. and Von Holle, B. (1999) Positive interactions of nonindigenous species: invasional meltdown? *Biological Invasions* 1, 21–32.

Stohlgren, T.J., Jarnevitch, C., Chong, G.W. and Evangelista, P.H. (2006) Scale and plant invasions: a theory of biotic acceptance. *Preslia* 78, 405–426.

Williamson, M. (1996) *Biological Invasions.* Chapman & Hall, London.

Williamson, M. and Brown, K.C. (1986) The analysis and modelling of British invasions. *Philosophical Transactions of the Royal Society* 314, 505–522.

yEd (2017) yEd Graph Editor. yWorks the diagramming company.

8 Biotic Resistance and Island Susceptibility Hypotheses

Jonathan M. Jeschke,[1,2,3]* Simon Debille[2,4] and Christopher J. Lortie[5]

[1]Leibniz-Institute of Freshwater Ecology and Inland Fisheries (IGB), Berlin, Germany; [2]Freie Universität Berlin, Institute of Biology, Berlin, Germany; [3]Berlin-Brandenburg Institute of Advanced Biodiversity Research (BBIB), Berlin, Germany; [4]KU Leuven – University of Leuven, Department of Biology, Leuven, Belgium; [5]York University, Department of Biology, Toronto, Canada

Abstract

The biotic resistance hypothesis *sensu stricto* is also known as the diversity–invasibility hypothesis. It proposes that ecosystems with high biodiversity are more resistant against non-native species than ecosystems with lower biodiversity. It is a classic hypothesis of the field and our systematic literature search identified 155 empirical studies that examined it. Most of these studies question the hypothesis. The frequency of supportive observational field studies is only about 15%. Although the frequency of supportive experimental studies, which are typically done at smaller spatial scales, is significantly higher, it is still below 50%. The island susceptibility hypothesis is topically similar and posits that continents are more resistant against non-native species than islands. In more specific terms, the island susceptibility hypothesis states that non-native species are more likely to become established and have major ecological impacts on islands than on continents. Our literature search only identified 17 empirical tests of this hypothesis with five of them being supportive. Thus, the biotic resistance and island susceptibility hypotheses are not frequently supported by existing empirical evidence. Most studies addressing them examined the number of non-native species or their establishment success, whereas relatively few studies measured impacts of non-native species. Studies that measured abundance, biomass or cover of non-native species – which are related to impact – more frequently supported the resistance hypothesis than other studies. A promising way forward might thus be to narrow the definition and scope of both hypotheses (and possibly rename them), so that 'resistance' and 'susceptibility' are related to impact of non-native species. The next steps will then be to critically test these revised hypotheses and further refine the relevant ecological contexts that mediate the importance or magnitude of resistance.

Introduction

Biotic resistance hypothesis

The biotic resistance hypothesis posits that ecosystems with high biodiversity are more resistant against non-native species than ecosystems with lower biodiversity (Elton,

* Corresponding author. E-mail: jonathan.jeschke@gmx.net

1958; Levine and D'Antonio, 1999). This hypothesis is tightly linked to Elton's definition of the ecological niche, according to which a niche is a characteristic of an ecosystem (Elton, 1927; see also Grinnell, 1917); this contrasts with Hutchinson's (1957) niche definition, which is a bit younger and more popular among today's ecologists. In Pulliam's (2000, p. 351) words: 'Hutchinson (1957) used the word niche to refer to the environmental requirements of a species, whereas earlier authors, especially Elton (1927) and Grinnell (1917), had used the term niche to refer to a place or "recess" in the environment that has the potential to support a species [...] According to Hutchinson, species, not environments, have niches.' These two definitions of the ecological niche are complementary and either can be useful.

Hence, according to Elton, environments and ecosystems have niches and these can be vacant. The basic idea underlying the resistance hypothesis is that ecosystems with high biodiversity have a low number of vacant niches (as most niches are filled by native species); consequently non-native species have few opportunities and resistance against them is high. Conversely, ecosystems with low biodiversity have a high number of vacant niches (they are not filled by native species); consequently non-native species have many opportunities and resistance against them is low. This reasoning seems to make sense intuitively, possibly because it resembles thoughts about market niches in economies where people talk about vacant niches as well. The reasoning highlights the importance of competition, an interaction where interaction partners are negatively affected and which has been a focus of ecologists for decades. The resistance hypothesis, however, basically ignores facilitation, an interaction where interacting partners are positively affected and which has only become a mainstream research topic more recently (Stachowicz, 2001; Bruno et al., 2003; Lortie and Callaway, 2009). This hypothesis also does not consider the characteristics of introduced non-native species (Jeschke et al., 2012a).

This chapter focuses on the biotic resistance hypothesis sensu stricto as defined above, which is also known as the diversity–invasibility hypothesis. A broader formulation – the biotic resistance hypothesis sensu lato – additionally includes the disturbance hypothesis, which is treated in the following chapter and focuses on reduced resistance owing to disturbance. Thinking even beyond the field of invasion biology, the biotic resistance hypothesis is a specific formulation of the diversity–stability hypothesis. This hypothesis in turn posits that ecosystems with high biodiversity are more stable than ecosystems with lower biodiversity (Ives and Carpenter, 2007; Jeschke et al., 2013; Jeschke, 2014). In the case of the resistance hypothesis, stability is specified to resistance against non-native species. Hence, the broader idea underlying the biotic resistance hypothesis is also very influential in other areas of ecological and biodiversity research. This hypothesis is thus a perfect example for one that tightly links invasion biology with other research fields (Jeschke, 2014).

Given this strong linkage with classic ecological research, it might not surprise that the biotic resistance hypothesis is one of the oldest invasion hypotheses. It was already featured in Elton's 1958 book The Ecology of Invasions by Animals and Plants that is often seen as the beginning of modern invasion biology, and only few invasion hypotheses date back even further (an example is Darwin's naturalization hypothesis featured in Chapter 15, this volume). The resistance hypothesis is very well known and reached the third place in a survey by Enders et al. (2018): only the enemy release (Chapter 11, this volume) and propagule pressure (Chapter 16, this volume) hypotheses were better known by the >350 invasion biologists taking this online survey. The empirical validity of the hypothesis has, however, been disputed, particularly for large spatial scales where contradictory findings have accumulated (e.g. Levine and D'Antonio, 1999; Stohlgren et al., 2003; Fridley et al., 2007; Jeschke et al., 2012a).

Island susceptibility hypothesis

The island susceptibility hypothesis is some-what similar to the biotic resistance hypothesis and is therefore treated in this chapter as well. In very general terms, it posits that continents are more resistant against non-native species than islands. More specifically, it says that non-native species are more likely to become established and have major ecological impacts on islands than on continents (Simberloff, 1995; Sol, 2000; Jeschke, 2008). Because continents typically have a higher biodiversity than islands (even when corrected for area), the two hypotheses treated in this chapter are logically linked to each other.

The island susceptibility hypothesis is not as important to the field as the resistance hypothesis, but it is also relatively well known among invasion biologists: in a survey by Enders *et al.* (2018) where participants could select from 33 presented invasion hypotheses those 1–3 hypotheses that they know best, only eight other invasion hypotheses were better known. This is probably owing to the fact that many well-known examples for high-impact invasions are from islands (e.g. Hawaiian islands, New Zealand's islands, Guam).

Goals of this chapter

After systematically reviewing the literature for both the biotic resistance hypothesis *sensu strico* and the island susceptibility hypothesis, we will apply the hierarchy-of-hypotheses (HoH) approach to structure currently available empirical tests, particularly for the relatively more common resistance hypothesis. We will thus address the following questions: (i) What aspects (i.e. sub-hypotheses) have been investigated thus far? (ii) Is there a geographic bias among studies on the biotic resistance and island susceptibility hypotheses? We will also apply a three-level ordinal scoring approach (Jeschke *et al.*, 2012a; Heger and Jeschke, 2014) in order to address the following questions: (iii) What is the level of support for the overall hypotheses and their sub-hypotheses? (iv) Does the level of support differ among major taxonomic groups, habitats, methodological approaches and over time?

Methods

Systematic literature search

As a basis for our current analysis, we used the empirical studies that we identified for a previous study (Jeschke *et al.*, 2012a) as relevant empirical tests of the biotic resistance hypothesis *sensu stricto* and the island susceptibility hypothesis. This search was done in the Web of Science on 19 February 2010, using the following string: '(biotic resistance OR resistance hypothesis OR diversity-invasibility hypothesis OR island susceptibility) AND (alien OR exotic OR introduced OR invasive OR naturali?ed OR nonindigenous OR non-native)'. We consulted the titles and abstracts of these articles and the full text of those that appeared potentially relevant. We also checked references cited in relevant articles. Purely theoretical tests of the hypotheses were not included, nor reviews or meta-analyses (these were excluded to avoid double-counting of empirical tests), but studies cited therein were included if relevant.

To update the dataset, we repeated the search on 2 July 2015 for both hypotheses (using the string given above). Because only few studies were identified for the island susceptibility hypothesis, we made an additional search for this hypothesis by inspecting papers citing Simberloff (1995), Sol (2000) and Jeschke (2008), which seem to be widely read publications in the context of this hypothesis. We searched for papers citing these publications in the Web of Science on 7 September 2016 and again consulted the titles and abstracts of these articles and the full text of those that appeared potentially relevant.

The updated dataset includes 155 empirical studies testing the biotic resistance hypothesis *sensu stricto* and 17 empirical studies testing the island susceptibility hypothesis. It is freely available online at www.hi-knowledge.org.

Hierarchy of hypotheses

The division of the resistance hypothesis into sub- and sub-sub-hypotheses was very similar to Jeschke *et al.* (2012a). We used the following categories to build the HoH:

1. Species richness: sub-hypothesis with native (or resident) species richness as measure of native (or resident) biodiversity, subdivided into the following sub-sub-hypotheses related to different inverse measures of resistance against non-native species: (i) establishment success of non-native species; (ii) number of non-native species (non-native species richness); (iii) abundance, biomass or cover of non-native species; (iv) survival, growth or reproduction of non-native species; and (v) other resistance measures (e.g. spread of non-native species); note that we followed Levine and D'Antonio (1999) and references cited therein in considering the percentage of non-native species as an inadequate inverse measure of resistance – this measure was thus not included in our HoH.
2. Shannon, evenness: sub-hypothesis with Shannon index or species evenness as measure of native (or resident) biodiversity.
3. Functional richness: sub-hypothesis with the number of functional groups among native (or resident) species as measure of native (or resident) biodiversity.
4. Proxy: sub-hypothesis with a proxy (e.g. latitude) as measure of native (or resident) biodiversity.

Sub-hypotheses 2, 3 and 4 could have been further divided into sub-sub-hypotheses as sub-hypothesis 1; however, we have not done so here due to the relatively low number of studies testing them. With more studies becoming available that test these sub-hypotheses, we recommend they should also be divided into sub-sub-hypotheses.

Similarly, because only relatively few studies have tested the island susceptibility

hypothesis until now, we decided not to divide it into sub-hypotheses but this can be done when more studies become available. Different measures of insularity (e.g. land-mass area or isolation) or susceptibility towards non-native species (e.g. establishment success and other inverse measures of resistance, see above) can be used for building an HoH of the island susceptibility hypothesis.

Scoring of empirical tests and analysis

We applied the three-level scoring approach as Jeschke *et al.* (2012a) and Heger and Jeschke (2014), i.e. we categorized the identified relevant empirical tests as either supporting, being undecided or questioning the biotic resistance and island susceptibility hypotheses. As outlined in Chapter 6, this volume, this scoring approach is different from vote counting which is only based on significance values and has key weaknesses. The scoring approach applied here takes all available evidence into account, particularly effect sizes, to classify studies as supporting, being undecided or questioning. These ordinal scores were used in the further analyses for which we used the statistical software program SPSS version 21. The dataset is freely available online at www.hi-knowledge. org.

Results

What aspects (i.e. sub-hypotheses) of the resistance hypothesis have been investigated thus far?

About 80% of the analysed studies investigating the resistance hypothesis measured native (or resident) biodiversity by recording the number of native (or resident) species (native species richness; n = 126 studies, Fig. 8.1). The remaining studies measured it by considering the number of functional groups among native (or resident) species (i.e. functional richness) or a few other estimates (Fig. 8.1). Due to this strong bias for using

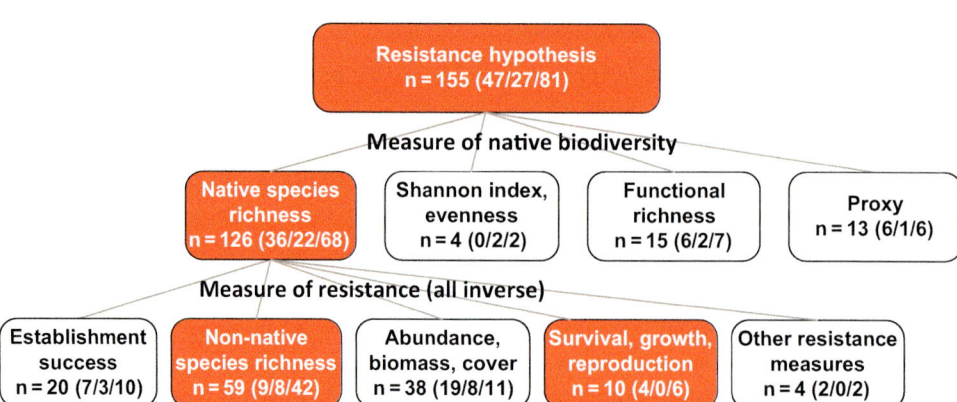

Fig. 8.1. The hierarchy of hypotheses for the biotic resistance hypothesis. The number of empirical studies related to each sub-(sub-)hypothesis add up to more than 155 studies because some studies are related to more than one sub-(sub-)hypothesis. The boxes are colour-coded: red indicates that >50% of the empirical studies are questioning the hypothesis, and n ≥5; green (not existent) would indicate that >50% of the empirical studies are supportive, and n ≥5; white is used for other cases (i.e. inconclusive data or n <5). Detailed information on the number of studies supporting, being undecided and questioning each (sub-sub-)hypothesis is provided in parentheses, e.g. for the overall hypothesis: 47 studies are supportive, 27 are undecided and 81 are questioning the resistance hypothesis.

species richness as measure of native biodiversity, only this sub-hypothesis was further divided into sub-sub-hypotheses. This further subdivision was done based on the (inverse) measure of resistance applied by each study. The most frequently applied one among the analysed studies was non-native species richness (n = 59 studies), followed by the abundance, biomass or cover of non-native species (n = 38), establishment success of non-native species (n = 20), survival, growth and reproduction of individuals of non-native species (n = 10) and other measures (n = 4). In other words, of the 20 sub-sub-hypotheses that can be built by the four outlined measures of native biodiversity and the five outlined measures of resistance, one sub-sub-hypothesis (native species richness as measure of native biodiversity combined with non-native species richness as inverse measure of resistance) was addressed by almost 40% of all analysed studies, and two sub-sub-hypotheses (the one just mentioned plus native species richness as measure of native biodiversity combined with abundance, biomass or cover of non-native species as inverse measure of resistance) were addressed by about 60% of all analysed

studies. This is a strong research bias towards testing a limited number of sub-sub-hypotheses.

Is there a geographic bias among studies on the biotic resistance and island susceptibility hypotheses?

We observed a strong geographic bias among existing studies on the resistance hypothesis: more than half of all studies were done in North America (n = 109), followed by Europe (n = 33), Australia/Oceania (n = 20), South America (n = 17), Asia (n = 15), Africa (n = 12) and Antarctica (n = 6). In contrast, most studies addressing the island susceptibility were global in scope, hence no strong differences in geographic scope were observed here.

What is the level of support for the hypotheses and sub-hypotheses?

Overall, the resistance hypothesis is relatively poorly supported by currently available studies: 53% (n = 81) of the 155 studies we analysed question this major hypothesis, 17% (n = 27) were undecided and 30% (n = 47) of all studies were supportive (Fig. 8.1). Taking a look at the sub-hypotheses, the most frequently addressed one is, of course, also questioned by empirical evidence because the empirical support for the overall hypothesis is mainly due to studies on this single sub-hypothesis. There are only few existing studies on the resistance hypothesis using the Shannon index or species evenness as measure of native biodiversity, and empirical support for the two remaining sub-hypotheses – functional richness and proxy measures of native biodiversity – was mixed (Fig. 8.1).

Interestingly, the levels of empirical support strongly differ between the two most frequently addressed sub-sub-hypotheses. Although the most frequently addressed one (native species richness as measure of native biodiversity combined with non-native species richness as inverse measure of resistance) is questioned by 42 of 59 studies (71%), the second-most frequently addressed one (native species richness as measure of native biodiversity combined with abundance, biomass or cover of non-native species as inverse measure of resistance) is only questioned by 11 of 38 studies (29%). The latter sub-sub-hypothesis is the one that has received most support of all (sub-)sub-hypotheses, with 50% of studies being supportive. Thus, none of the (sub-)sub-hypotheses of the resistance hypothesis has received more than 50% support.

Similarly, the island susceptibility hypothesis has received low empirical support. Of the 17 studies addressing it, five studies (29%) provided supporting evidence, four (24%) were undecided and eight (47%) were questioning the hypothesis. Thirteen of these studies used establishment success of non-native species as measure of susceptibility, two studies used the number of non-native species per plot; and the two remaining studies measures related to impact.

Does the level of support differ among major taxonomic groups, habitats, methodological approaches and over time?

Most studies (73%) on the biotic resistance hypothesis have been done on non-native plants, whereas comparatively few studies are available on non-native invertebrates (13%) and vertebrates (14%; Fig. 8.2a). There are no statistically significant differences in the level of support among taxonomic groups (Mann–Whitney U-tests, all two-sided; the same is true for all other significance tests in this chapter): plants vs invertebrates, $p = 0.80$; plants vs vertebrates, $p = 0.24$; invertebrates vs vertebrates, $p = 0.31$).

There was an even stronger research bias when comparing major environmental habitats: by far most studies have been performed in terrestrial habitats (85%), whereas freshwater and marine habitats are understudied (9% and 7% of the available tests, respectively; Fig. 8.2b). Thus, most studies addressing the resistance hypothesis focus on terrestrial plants (n=109, or 70% of the 155 studies in total on this hypothesis). Studies in terrestrial habitats showed a significantly lower level of empirical support than studies in aquatic habitats (freshwater and marine combined, $p = 0.019$, U-test, Fig. 8.2b); individual comparisons showed a marginally significant difference between terrestrial and freshwater habitats ($p = 0.053$; terrestrial vs marine, $p = 0.13$; freshwater vs marine, $p = 0.98$; U-tests).

We separated the studies according to their methodological approach into: (i) laboratory studies that were all experimental; (ii) experimental field and enclosure (incl. exclosure) studies; and (iii) observational or correlational studies that were all done in the field. Although laboratory studies on the resistance hypothesis are currently rare, there are similar numbers of experimental field and enclosure studies, on the one hand, and observational field studies, on the other hand (Fig. 8.2c). The level of empirical support tended to decline in this direction (Fig. 8.2c), where experimental

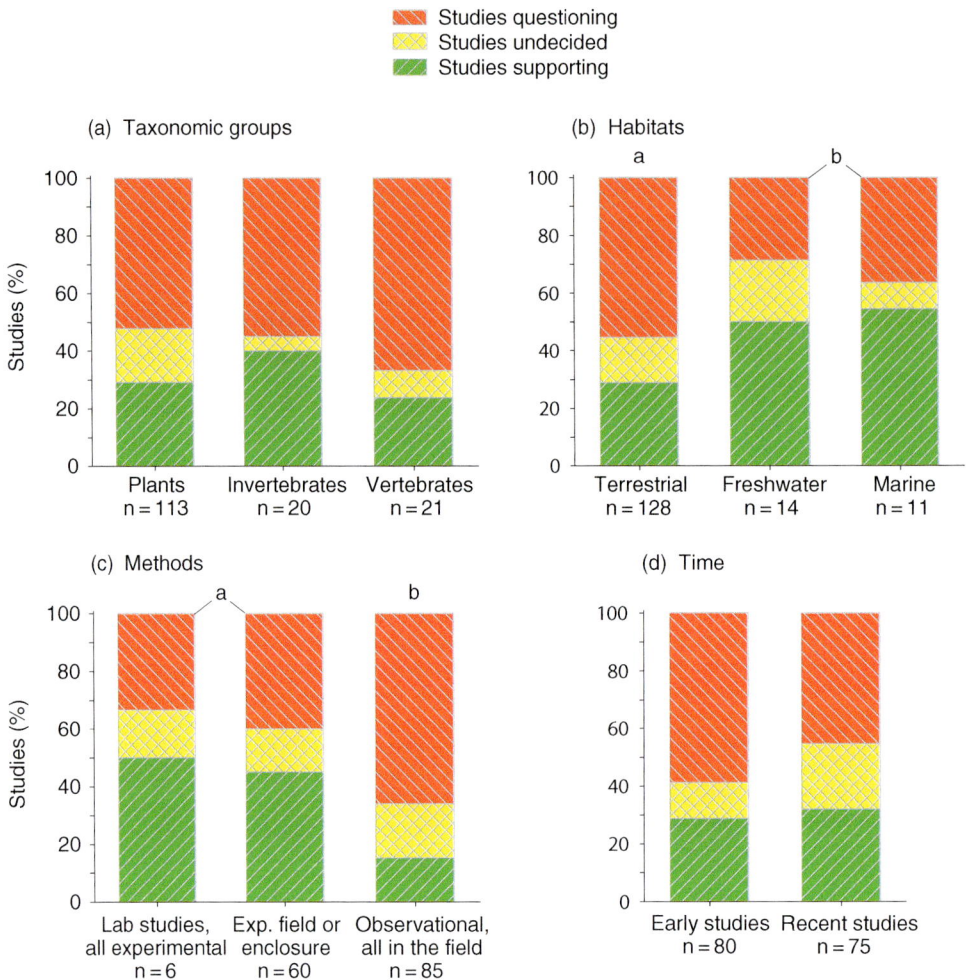

Fig. 8.2. Empirical level of support for the biotic resistance hypothesis, subdivided for (a) major taxonomic groups, (b) major habitat types (here, the number of studies do not add up to 155 because studies in multiple habitats were excluded from this comparison), (c) methodological approaches (again, studies using multiple methods were excluded from this comparison) and (d) early vs recent studies. Letters indicate statistically significant differences (U-tests, $p < 0.05$).

studies showed a similar level of support that was significantly higher than the level shown by observational studies ($p < 0.001$; due to the small sample size of experimental laboratory studies, we combined them with other experimental studies for this U-test).

We divided the studies in our dataset between early studies published until 2006 and recent studies published thereafter, using the cut-off year 2006 to be as close as possible to a 50:50 division between early and recent studies (cf. Jeschke *et al.*, 2012a). There was no statistically significant difference in the level of support between early and recent studies ($p = 0.20$, U-test).

In case of the island susceptibility hypothesis, more studies are currently available on vertebrates than plants, whereas no study in our dataset focused on invertebrates (Fig. 8.3a). There were no strong differences in empirical support between studies on vertebrates vs plants ($p = 0.63$, U-test). Most studies addressing the island susceptibility hypothesis were done in terrestrial habitats and a few in freshwaters (Fig. 8.3b). Methodologically, all papers we identified addressing this hypothesis were observational field studies. Finally, there were no statistically significant differences in the level of empirical support between early studies (published until 2006) and recent studies (published thereafter; 2006 was again the cut-off year to achieve a circa 50:50 division between early and recent studies) (Fig. 8.3c; $p = 0.35$, U-test).

Discussion

What have we learned?

Only a relatively small proportion of available empirical studies support the biotic resistance and the island susceptibility hypothesis. This finding is in line with previous studies (e.g. Levine and D'Antonio, 1999; Fridley *et al.*, 2007; Jeschke *et al.*, 2012a). Applying the HoH approach, we showed strong biases in the research coverage of the resistance hypothesis' (sub-)sub-hypotheses: one of four sub-hypotheses and two of 20 sub-sub-hypotheses that we identified have been frequently addressed by

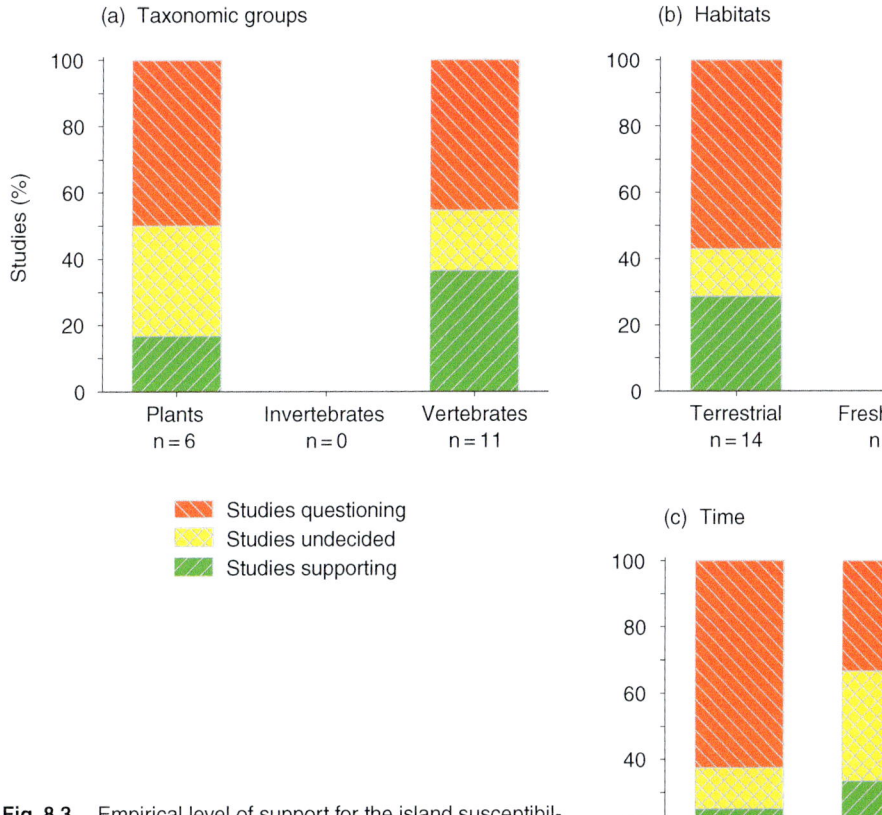

Fig. 8.3. Empirical level of support for the island susceptibility hypothesis, subdivided for (a) major taxonomic groups, (b) major habitat types and (c) early vs recent studies. Missing bars indicate lacking data: bars are only shown if at least five studies were carried out in a given category. There were no statistically significant differences.

empirical studies but other (sub-)sub-hypotheses only relatively rarely. There are profound differences in the level of support among sub-sub-hypotheses. In particular, the sub-sub-hypothesis with the relative majority of studies (n = 59) has native species richness as measure of native biodiversity and non-native species richness as inverse measure of resistance but only 15% of these studies were supportive. The sub-sub-hypothesis that has been addressed by the second highest number of studies (n = 38) also has native species richness as measure of native biodiversity but abundance, biomass or cover of non-native species as inverse measure of resistance; 50% of these studies were supportive. This sub-sub-hypothesis was the one with the highest level of empirical support among all (sub-)sub-hypotheses of the biotic resistance hypothesis. We will follow up on this result in the next section where we suggest narrowing the definition and scope of the resistance hypothesis, so that 'resistance' is related to impact of non-native species.

Another interesting finding is that experimental studies showed a significantly higher level of empirical support than observational studies. Although we did not extract specific information on the spatial scale of each study, it seems clear that the experimental studies in our dataset, which were done in laboratories, enclosures or in the field, were done at smaller spatial scales on average than the observational studies, which were all done in the field. Thus, the difference in empirical support we observed is in line with earlier publications noting a difference in empirical support between studies done at small vs large spatial scales (Fridley *et al.*, 2007).

Most studies on the resistance and island susceptibility hypotheses were done in terrestrial habitats. In the case of the resistance hypothesis, these studies showed particularly low levels of empirical support. One reason might be that community turnover in terrestrial systems dominated by long-lived plants is often slower than in aquatic systems. Consequently, time lags in colonization, spread and abundance can be more pronounced in terrestrial systems, possibly masking potential effects of diversity. This is highly speculative, though, and should be scrutinized in the future. In the case of the island susceptibility hypothesis, too few non-terrestrial studies are available for a meaningful comparison. Plants predominate as focal non-native species in tests of the resistance hypothesis, whereas more vertebrate studies exist for the island susceptibility hypothesis (cf. Jeschke *et al.*, 2012b). We did not observe strong differences in empirical support among taxonomic groups for either of the two hypotheses. These findings are in line with Jeschke *et al.* (2012a), which was based on a slightly smaller dataset.

When comparing the results reported here with those of Jeschke *et al.* (2012a), it should be considered that each paper was only counted once for each comparison and at each level of the HoH here, which is why the number of studies in Fig. 8.1 differs among the hierarchical levels: summing up the number of studies across all (sub-)sub-hypotheses yields n = 163 studies because a few studies addressed multiple (sub-)sub-hypotheses. These were combined at higher levels of the HoH so that each study is only counted once, yielding n = 155 studies (i.e. publications) in total. In Jeschke *et al.* (2012a), however, those (few) papers that looked at two or more sub-hypotheses were included two or more times in the HoH, hence they were not combined at higher levels. Thus, we followed a more conservative approach here, resulting in slightly smaller sample sizes. The number of studies is still higher than in Jeschke *et al.* (2012a) owing to the updated and thus larger dataset.

The strong geographic bias we observed for the resistance hypothesis is in line with previous analysis about research on biological invasions (Pyšek *et al.*, 2008; Bellard and Jeschke, 2016). Thus, the low geographic bias for the island susceptibility hypothesis is relatively unusual for the field and due to the high number of global studies on this hypothesis.

In contrast to Jeschke *et al.* (2012a), we did not observe a decline in empirical support of the two hypotheses over time. Both hypotheses have, however, received only little support overall.

How can we move on?

The low levels of empirical support that we and others observed for the two hypotheses suggest we should consider revising or even abandoning them. It could be that these hypotheses simply do not work, for instance because they: (i) focus on competition between species and ignore facilitation; and (ii) ignore characteristics of non-native species (see the Introduction section). It seems too early to completely abandon them, however, and revising them seems more appropriate. A closer look at our and earlier results reveals possibilities for doing so.

First, and as pointed out in previous publications (e.g. Fridley *et al.*, 2007), the resistance hypothesis is better empirically supported at small rather than large spatial scales. It would thus be possible to restrict its application to small spatial scales and abandon it when focusing on large scales.

Second, our results show that most studies addressing these hypotheses looked at numbers of non-native species (in the case of the resistance hypothesis) or their establishment success (in the case of the island susceptibility hypothesis), whereas relatively few studies have looked at impacts of non-native species. Those studies addressing the resistance hypothesis that looked at abundance, biomass or cover of non-native species – which are more closely related to impact than species numbers – showed higher levels of support than other studies on this hypothesis. Levine *et al.* (2004) similarly concluded from their results 'that ecological interactions rarely enable communities to resist invasion, but instead constrain the abundance of invasive species once they have successfully established'. In the case of the island susceptibility hypothesis, only two studies looked at impact *sensu lato*. This sample size is too low to be

meaningful but at least neither of the two studies questioned the hypothesis.

It does not surprise that most currently available studies addressing the two hypotheses focused on numbers of non-native species or establishment success because these were the classic quantities to look at in invasion biology. Impacts (and pathways) have only recently become core and mainstream research topics (Jeschke *et al.*, 2014; Kumschick *et al.*, 2015). Only now has the field reached a stage where impact assessments are done with a large coverage of species, and impact assessments are on the way to being implemented in large invasion databases such as the IUCN Global Invasive Species Database (GISD, www.iucngisd.org) using the Environmental Impact Classification for Alien Taxa, EICAT (Blackburn *et al.*, 2014; Hawkins *et al.*, 2015).

A promising way forward might be to narrow the definition and scope of the resistance and island susceptibility hypotheses, so that 'resistance' and 'susceptibility' are related to impact of non-native species. This would be highly useful and valuable, given the crucial importance of understanding invader impacts (Jeschke *et al.*, 2014; Kumschick *et al.*, 2015). Specifically, the biotic resistance hypothesis could be revised as follows: ecosystems with high biodiversity are more resistant against non-native species than ecosystems with lower biodiversity, leading to lower levels of impact in highly diverse systems. This narrower definition of the resistance hypothesis could also be called *impact resistance hypothesis*. Similarly, the island susceptibility hypothesis could be revised as follows: non-native species are more likely to have ecological impacts on islands than on continents. This narrower hypothesis could also be called *island impact hypothesis*.

The next step will then be to critically test these revised hypotheses. For instance, EICAT can be used to assess impact, and assessment results can be compared to: (i) biodiversity in order to address the diversity-impact hypothesis; and (ii) insularity in order to address the island impact hypothesis.

References

Bellard, C. and Jeschke, J.M. (2016) A spatial mismatch between invader impacts and research publications. *Conservation Biology* 30, 230–232.

Blackburn, T.M., Essl, F., Evans, T., Hulme, P.E., Jeschke, J.M. *et al.* (2014) A unified classification of alien species based on the magnitude of their environmental impacts. *PLoS Biology* 12, e1001850.

Bruno, J.F., Stachowicz, J.J. and Bertness, M.D. (2003) Inclusion of facilitation into ecological theory. *Trends in Ecology & Evolution* 18, 119–125.

Elton, C.S. (1927) *Animal Ecology.* Sidgwick & Jackson, London.

Elton, C.S. (1958) *The Ecology of Invasions by Animals and Plants.* Methuen, London.

Enders, M., Hütt, M.-T. and Jeschke, J.M. (2018) Drawing a map of invasion biology based on a network of hypotheses. *Ecosphere.* DOI: 10.1002/ecs2.2146.

Fridley, J.D., Stachowicz, J.J., Naeem, S., Sax, D.F., Seabloom, E.W. *et al.* (2007) The invasion paradox: reconciling pattern and process in species invasions. *Ecology* 88, 3–17.

Grinnell, J. (1917) The niche-relationships of the California thrasher. *Auk* 34, 427–433.

Hawkins, C.L., Bacher, S., Essl, F., Hulme, P.E., Jeschke, J.M. *et al.* (2015) Framework and guidelines for implementing the proposed IUCN Environmental Impact Classification for Alien Taxa (EICAT). *Diversity and Distributions* 21, 1360–1363.

Heger, T. and Jeschke, J.M. (2014) The enemy release hypothesis as a hierarchy of hypotheses. *Oikos* 123, 741–750.

Hutchinson, G.E. (1957) Concluding remarks. *Cold Spring Harbor Symposia on Quantitative Biology* 22, 415–427.

Ives, A.R. and Carpenter, S.R. (2007) Stability and diversity of ecosystems. *Science* 317, 58–62.

Jeschke, J.M. (2008) Across islands and continents, mammals are more successful invaders than birds. *Diversity and Distributions* 14, 913–916.

Jeschke, J.M. (2014) General hypotheses in invasion ecology. *Diversity and Distributions* 20, 1229–1234.

Jeschke, J.M., Gómez Aparicio, L., Haider, S., Heger, T., Lortie, C.J., Pyšek, P. and Strayer, D.L. (2012a) Support for major hypotheses in invasion biology is uneven and declining. *NeoBiota* 14, 1–20.

Jeschke, J.M., Gómez Aparicio, L., Haider, S., Heger, T., Lortie, C.J., Pyšek, P. and Strayer, D.L. (2012b) Taxonomic bias and lack of cross-taxonomic studies in invasion biology. *Frontiers in Ecology and the Environment* 10, 349–350.

Jeschke, J. M., Keesing, F. and Ostfeld, R. S. (2013) Novel organisms: comparing invasive species, GMOs, and emerging pathogens. *Ambio* 42, 541–548.

Jeschke, J.M., Bacher, S., Blackburn, T.M., Dick, J.T.A., Essl, F. *et al.* (2014) Defining the impact of non-native species. *Conservation Biology* 28, 1188–1194.

Kumschick, S., Gaertner, M., Vilà, M., Essl, F., Jeschke, J.M. *et al.* (2015) Ecological impacts of alien species: quantification, scope, caveats, and recommendations. *BioScience* 65, 55–63.

Levine, J.M. and D'Antonio, C.M. (1999) Elton revisited: a review of evidence linking diversity and invasibility. *Oikos*, 87, 15–26.

Levine, J.M., Adler, P.B. and Yelenik, S.G. (2004) A meta-analysis of biotic resistance to exotic plant invasions. *Ecology Letters* 7, 975–989.

Lortie, C.J. and Callaway, R.M. (2009) David and Goliath: comparative use of facilitation and competition studies in the plant ecology literature. *Web Ecology* 9, 54–57.

Pulliam, H. R. (2000) On the relationship between niche and distribution. *Ecology Letters* 3, 349–361.

Pyšek, P., Richardson, D.M., Pergl, J., Jarošík, V., Sixtová, Z. and Weber, E. (2008) Geographical and taxonomic biases in invasion ecology. *Trends in Ecology & Evolution* 23, 237–244.

Simberloff, D. (1995) Why do introduced species appear to devastate islands more than mainland areas? *Pacific Science* 49, 87–97.

Sol, D. (2000) Are islands more susceptible to be invaded than continents? Birds say no. *Ecography* 23, 687–692.

Stachowicz, J.J. (2001) Mutualism, facilitation, and the structure of ecological communities. *BioScience* 51, 235–246.

Stohlgren, T.J., Barnett, D. and Kartesz, J.T. (2003) The rich get richer: patterns of plant invasions in the United States. *Frontiers in Ecology and the Environment* 1, 11–14.

9

Disturbance Hypothesis

Regina Nordheimer[1] and Jonathan M. Jeschke[1,2,3]*

[1]Freie Universität Berlin, Institute of Biology, Berlin, Germany; [2]Leibniz-Institute of Freshwater Ecology and Inland Fisheries (IGB), Berlin, Germany; [3]Berlin-Brandenburg Institute of Advanced Biodiversity Research (BBIB), Berlin, Germany

Abstract

The disturbance hypothesis posits that the invasion success of non-native species is higher in highly disturbed than in relatively undisturbed ecosystems. A synonymous formulation is that highly disturbed ecosystems show lower resistance against non-native species than relatively undisturbed ecosystems. On the basis of a systematic literature search, we identified 126 studies addressing the disturbance hypothesis. Applying the hierarchy-of-hypotheses approach, we classified these studies according to: (i) the cause of disturbance (direct anthropogenic vs indirect or non-anthropogenic disturbances); (ii) the type of disturbance; and (iii) the measure of invasion success. The majority of studies reported evidence supporting the disturbance hypothesis (59%, with 21% of the studies questioning the hypothesis and another 21% being undecided). Most sub-hypotheses were supported as well, only studies focusing on conservation management or fire as disturbances frequently reported questioning evidence. There was also consistent support for the hypothesis across taxonomic groups and terrestrial as well as aquatic systems. However, experimental field studies showed a lower level of support than observational or laboratory studies. Overall, this hypothesis is relatively well, but not very strongly supported by currently available studies. Given that human disturbances will further increase in the foreseeable future, we can predict that these will tend to further promote biological invasions.

Introduction

The disturbance hypothesis posits that the invasion success of non-native species is higher in highly disturbed than in relatively undisturbed ecosystems. A synonymous formulation is that highly disturbed ecosystems show lower resistance against non-native species than relatively undisturbed ecosystems. The basic idea underlying this hypothesis is that pre-existing conditions typically favour native and disfavour non-native species because the native species are adapted to these conditions. Consequently, non-native species should benefit if the conditions are disturbed.

Disturbance has been defined in various ways by different ecologists, but many if not most definitions refer to Grime (1977) or White and Pickett (1985). Grime (1977) wrote that disturbance 'is associated with the partial or total destruction of the plant biomass and arises from the activities of herbivores, pathogens, man (trampling, mowing, and ploughing), and from phenomena such as wind damage, frosts, desiccation, soil erosion, and fire' (p. 1169). This classic definition focuses on plants, whereas White and

* Corresponding author. E-mail: jonathan.jeschke@gmx.net

Pickett's (1985) definition is more general: 'a disturbance is any relatively discrete event in time that disrupts ecosystem, community, or population structure and changes resources, substrate availability, or the physical environment' (p. 7). This definition forms the basis of how we specify disturbance for the purposes of this chapter. We would like to highlight that we do not restrict the term to short and abrupt events but include long-lasting processes and changes such as urbanization. In this chapter, a disturbance is defined as an event or process that disrupts ecosystem, community or population structure, and changes resources, substrate availability or the physical environment. Disturbances can have anthropogenic or non-anthropogenic causes (Sousa, 1984). We excluded disturbances due to non-native species in the wild in order to avoid overlap with the similar invasional meltdown hypothesis (see the following Chapter 10), but did include disturbances due to livestock. In summary, we focus on disturbances that are directly or indirectly caused by humans (including effects of livestock, with the exception of the effects of non-native species in the wild) or occur as natural phenomena such as fire, storms or floods.

The disturbance hypothesis is another classic one and is as old as the biotic resistance hypothesis featured in the previous chapter (dating back to Elton, 1958; for the disturbance hypothesis, see also e.g. Rejmánek, 1989; Hobbs and Huenneke, 1992; Sher and Hyatt, 1999; Moles et al., 2012). These hypotheses are somewhat similar because both are related to resistance against non-native species. Disturbance can lead to lower biodiversity, adding an additional link between the two hypotheses. Furthermore, many disturbances are anthropogenic; hence the disturbance hypothesis is related to the propagule pressure hypothesis (see Chapter 16, this volume) in that both of these hypotheses highlight the role of humans and their actions as key drivers of biological invasions. However, the hypothesis that is most similar to the disturbance hypothesis is arguably the invasional meltdown hypothesis, which was already mentioned above, because

invaders can have strong impacts and thus disturb ecosystems and communities.

The disturbance hypothesis is a very influential and popular invasion hypothesis. It had the second-highest number of studies focusing on it in the systematic review by Lowry et al. (2013). In a recent survey among >350 experts, it was the third best-known invasion hypothesis (Enders et al., 2018).

However, to which degree this hypothesis is empirically supported has been unclear. We fully agree with Moles et al. (2012) who wrote: 'Several foundational ideas in invasion biology have become widely accepted without appropriate testing, or despite equivocal evidence from empirical tests. One such idea is the suggestion that disturbance facilitates invasion.' Their analysis of plant data from 200 sites across the world suggested that disturbance is only a weak predictor of invasion. Many studies have, however, provided supporting evidence for the disturbance hypothesis (e.g. Burke and Grime, 1996; Alston and Richardson, 2006; Johnson et al., 2008; Malumbres-Olarte et al., 2014).

To shed more light on the disturbance hypothesis and its level of empirical support, we here combine the hierarchy-of-hypotheses (HoH) approach with a systematic literature review across taxonomic groups, applying a three-level ordinal scoring approach. Specifically, we address the following questions: (i) Which aspects (i.e. sub-hypotheses) of the disturbance hypothesis have been investigated thus far? (ii) What is the level of support for the overall hypothesis and its sub-hypotheses? (iii) Does the level of support differ among major taxonomic groups, habitats and methodological approaches? (iv) Has the level of support changed over time (cf. Jeschke et al., 2012a)?

Methods

Systematic literature search

We searched the ISI Web of Science on 11 September 2015, using the following string: 'Disturb* AND hypothes* AND (alien OR exotic OR introduc* OR invas* OR naturali?ed OR nonindigenous OR non-indigenous OR nonnative OR non-native)'. This search returned 1295 hits that we screened by title and abstract. We consulted the full text of those 524 articles that appeared potentially relevant and additionally considered articles cited in relevant studies. In this way, we identified 126 relevant studies: those that provide empirical data addressing the disturbance hypothesis. Purely theoretical tests of the hypothesis were not included, nor were reviews or meta-analyses (these were excluded to avoid double-counting of empirical tests).

Hierarchy of hypotheses

Relevant studies were divided into sub-, sub-sub- and lowest-level hypotheses according to the cause of disturbance, type of disturbance and measure of invasion success. We thereby discriminated two different causes of disturbance (direct anthropogenic vs indirect or non-anthropogenic disturbances) and differentiated several types of disturbance for each of these causes:

1. Direct anthropogenic disturbances
   a. Urbanization: proximity to urban areas, other human settlements or streets; pollution due to urbanization or industry (studies better matching the following category, 'cultivation', were put into that category).
   b. Cultivation and economic land use: disturbances as part of land-use practices such as fertilization, mowing, soil turnover with a spade or grazing as part of agricultural land use; forest regeneration and fragmentation or timber production.
   c. Management for conservation or restoration purposes: disturbances for other reasons than cultivation and land use due to management actions, e.g. use of herbicides and active fire management (otherwise, fire is included in 2a below), mowing or ploughing as management actions for conservation or restoration purposes (otherwise,

mowing is included in 1b); studies that focused on managing the focal non-native species were not included, only those where the management actions included natives as target species.
   d. Other types of direct anthropogenic disturbance that could not be included in the above categories: e.g. disturbance by visitors in national parks, use of military vehicles or due to artificially constructed lakes and ponds.
2. Indirect or non-anthropogenic disturbances
   a. Fire: all types of spontaneous fire (not intentionally induced by humans).
   b. Flood or storm, including related changes such as uprooted trees.
   c. Non-anthropogenic soil disturbance, e.g. by wild boar or hedgehogs.
   d. Global warming, including increased CO_2 concentration.

In addition, we discriminated the following measures of invasion success that studies related to the level of disturbance in order to test the hypothesis:

1. Number of exotic species (exotic species richness).
2. Establishment or colonization success of exotic species (proportion of introduced species that successfully established themselves or colonized the exotic range).
3. Occurrence of exotic species (presence/absence).
4. Abundance, biomass or cover of exotic species.
5. Survival or recruitment/reproduction of exotic individuals.
6. Growth or emergence of exotic individuals.

Scoring of empirical tests and analysis

We applied the three-level scoring approach as described in Jeschke et al. (2012a) and Heger and Jeschke (2014), i.e. we categorized the identified relevant empirical tests as either supporting, being undecided or questioning the disturbance hypothesis. As outlined in Chapter 6, this volume, this approach differs from vote counting, which is only based on significance values and has key weaknesses. The approach applied here takes all available evidence into account, particularly effect sizes, to classify studies as supporting, being undecided or questioning. These ordinal scores were used in the further analyses for which we used the statistical software program SPSS version 21. The dataset is freely available online at www.hi-knowledge.org.

Results

Which aspects (i.e. sub-hypotheses) of the disturbance hypothesis have been investigated thus far?

Given our definition of disturbance, we identified more studies focusing on direct anthropogenic rather than other causes of disturbance ($n = 96$ vs $n = 32$). Within the category of direct anthropogenic disturbances, most studies focused on urbanization and economic land use (Fig. 9.1). At the lowest level of the hierarchy of hypotheses, the most frequently used measures of invasion success were abundance, biomass or cover of exotics and number of exotic species. Of the 48 lowest-level hypotheses, 18 were not addressed by any of the 126 studies in our dataset and 21 were addressed by <5 studies. Thus, only nine lowest-level hypotheses were addressed by at least five studies. The most frequently investigated ones were, in this order (all direct anthropogenic disturbances): economic land use as type of disturbance combined with abundance, biomass or cover as measure of invasion success ($n = 25$); urbanization as type of disturbance combined with abundance, biomass or cover as measure of invasion success ($n = 18$); and urbanization as type of disturbance combined with number of exotic species as measure of invasion success ($n = 17$; Fig. 9.1).

There was a strong geographic bias among the studies in the dataset: more than 50% of them focused on North America ($n = 67$), whereas fewer than 15% of studies focused on any of the other continents

(Europe: $n = 18$; Australia/Oceania: $n = 16$; South America: $n = 14$; Africa: $n = 7$; Asia: $n = 4$; Antarctica: $n = 0$).

What is the level of support for the overall hypothesis and its sub-hypotheses?

The majority of the 126 studies in the dataset reported evidence supporting the disturbance hypothesis (59%, with 21% of the studies questioning the hypothesis and another 21% being undecided; Fig. 9.1). Both sub-hypotheses – direct anthropogenic vs indirect or non-anthropogenic cause of disturbance – were also supported by the majority of studies. Regarding the type of disturbance, five of the seven sub-sub-hypotheses with at least five studies were supported by the majority of studies as well. This was not the case for conservation management and fire (Fig. 9.1). At the lowest and most detailed level of the hierarchy of hypotheses, seven of the nine lowest-level hypotheses that were addressed by at least five studies were again supported by the majority of studies. More studies were questioning than supporting the two remaining lowest-level hypotheses: conservation management as type of disturbance combined with abundance, biomass or cover of exotics; and fire combined with abundance, biomass or cover of exotics (Fig. 9.1).

Does the level of support differ among major taxonomic groups, habitats and methodological approaches?

The majority of studies supported the disturbance hypothesis across major taxonomic groups and habitat types. Although there was a slight tendency that support was higher among studies focusing on vertebrates or invertebrates than on plants, these differences were not statistically significant (Mann–Whitney U-tests, all two-sided (the same is true for all other significance tests in this chapter): plants vs invertebrates, $p = 0.51$; plants vs vertebrates, $p = 0.38$; invertebrates vs vertebrates, $p = 0.77$;

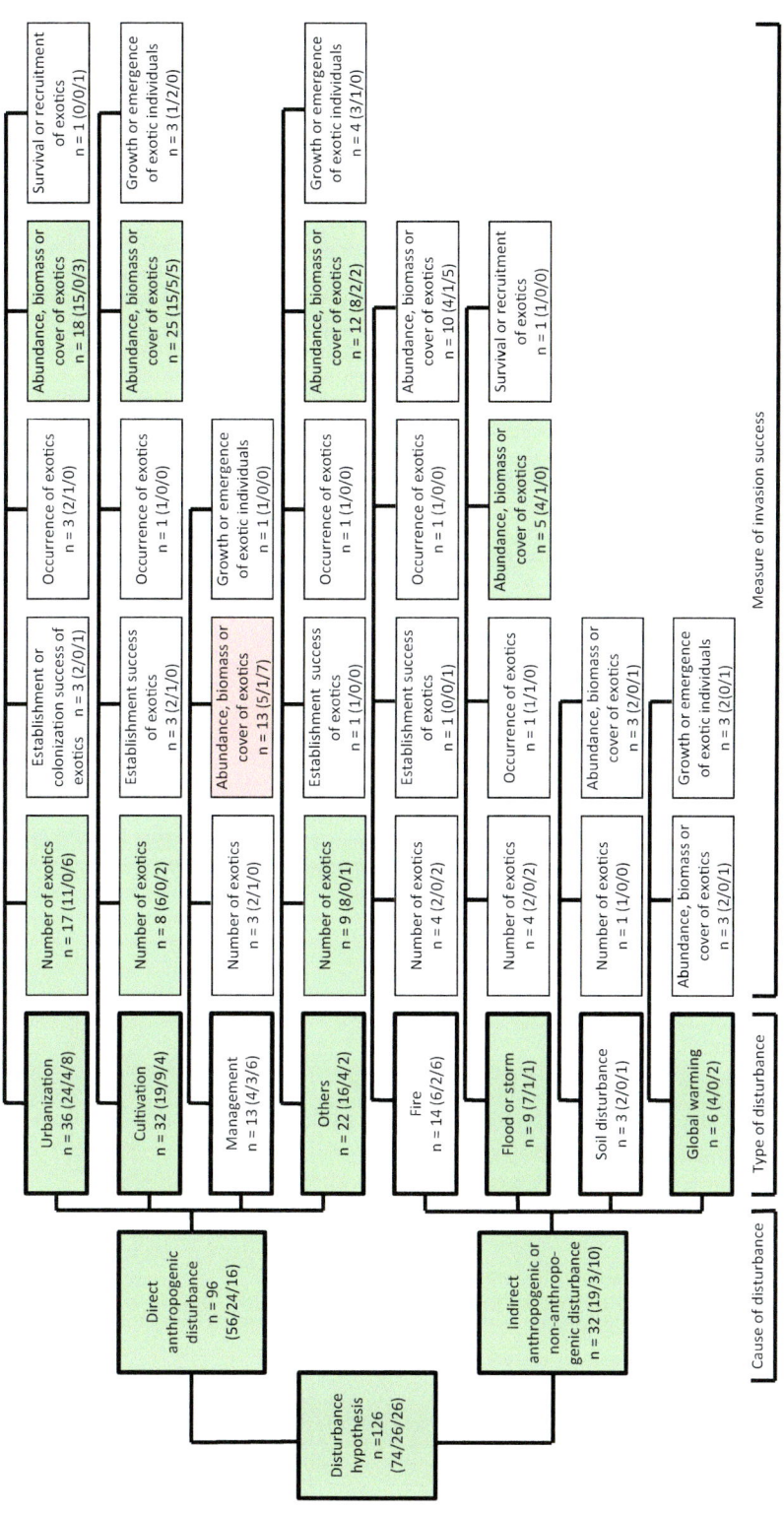

Fig. 9.1. The hierarchy of hypotheses for the disturbance hypothesis. The number of empirical studies related to each sub-hypothesis adds up to more than 126 studies because some studies are related to more than one sub-hypothesis. The boxes are colour coded: green indicates that >50% of the empirical studies are supportive, and $n \geq 5$; red indicates that >50% of the empirical studies are questioning the hypothesis, and $n \geq 5$; white is used for other cases (i.e. inconclusive data or $n < 5$). Detailed information on the number of studies supporting, being undecided and questioning each (sub-)hypothesis is provided in parentheses, e.g. for the overall hypothesis: 74 studies are supportive, 26 are undecided and 26 are questioning the resistance hypothesis.

Fig. 9.2). Many more studies have been carried out on plants rather than vertebrates and invertebrates combined. Similarly, many more studies have been done in terrestrial rather than freshwater or marine habitats. Empirical support for the disturbance hypotheses was similar for these three habitat types (Fig. 9.2). Thus, most studies have been carried out on terrestrial plants (n = 91, i.e. 72%), with only relatively small numbers of studies having investigated other taxa and habitats. Combining this bias regarding

taxonomic group and habitat with the geographic bias reported above, we can conclude that the relative majority of studies have focused on terrestrial plants in North America (n = 57, i.e. 45%).

Interestingly, studies applying the arguably strongest methodological approach, experiments in the field or enclosures, reported the lowest level of empirical support: here, only 42% of the studies were supportive (n = 20), 31% reported questioning evidence (n = 15) and 27% were

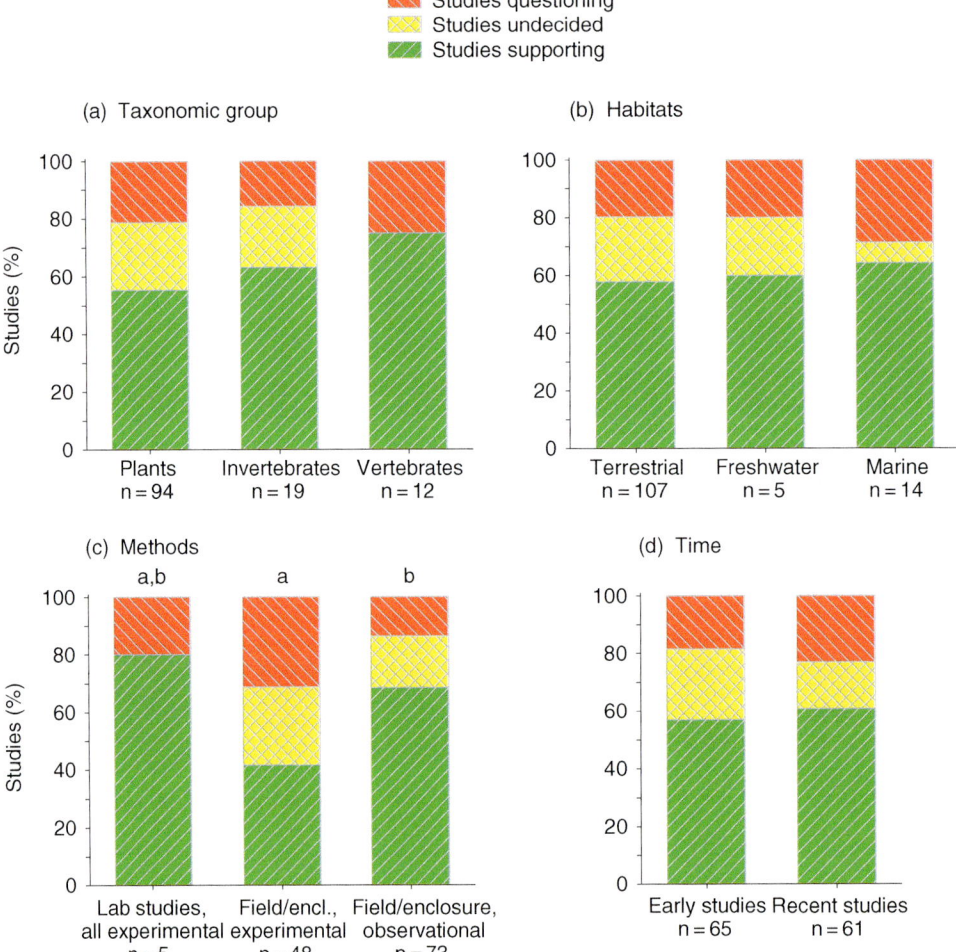

Fig. 9.2. Empirical level of support for the disturbance hypothesis, subdivided for (a) major taxonomic groups (here, the numbers do not add up to 126 studies because not all taxonomic groups are shown, e.g. algae are missing), (b) major habitat types, (c) methodological approaches and (d) early vs recent studies. Letters in panel c indicate statistically significant differences (U-tests, $p < 0.05$). No statistically significant differences were observed for panels a, b and d.

undecided (n = 13; Fig. 9.2). Observational field or enclosure studies had a significantly higher level of empirical support (n = 50 being supportive, n = 10 questioning, n = 13 being undecided; $p = 0.003$). There were only few laboratory studies, hence it may not surprise that differences between these and field/enclosure studies were not statistically significant (lab vs experimental field or enclosure studies, $p = 0.24$; lab vs observational field or enclosure studies, $p = 0.78$).

Has the level of support changed over time?

We divided the empirical studies in our dataset between early studies published until 2006 and recent studies published thereafter, using the cut-off year 2006 to be as close as possible to a 50:50 division between early and recent studies (cf. Jeschke et al., 2012a). There was no apparent change in the level of support for the disturbance hypothesis over time (Fig. 9.2; $p = 0.90$).

Discussion

Overall, the disturbance hypothesis has been largely empirically supported, including most of its sub-hypotheses and across taxonomic groups and habitats. The only 'disturbing' finding was that studies with the strongest methodological approach (experimental field and enclosure studies) have been ambiguous: here, the hypothesis is not well supported although still more studies reported supporting than questioning evidence. One explanation could be that a recognizable fraction of these studies focused on conservation management or fire as disturbances, which, according to our dataset, generally have lower levels of empirical support. Interestingly, although studies on conservation management did not include studies that directly targeted the non-native species, the actions apparently indirectly reduced the abundance, biomass or cover of non-natives.

A temporal effect was observed in some of the studies investigating fire: directly after the disturbance, the invasion success of non-native species tended to increase, thus supporting the disturbance hypothesis; however, this effect diminished with time (Dodge et al., 2008; Shive et al., 2013). A similar effect was observed for human trampling as a disturbance (Hernandez and Sandquist, 2011). More research is needed to clarify such temporal effects but a challenge is that some types of disturbance cannot, or should not, be experimentally induced in the field (e.g. storms).

Studies on the disturbance hypothesis show strong biases regarding geography, taxonomy and the focal type of habitat. This comes as no surprise, though, because similar biases were reported by Jeschke et al. (2012a,b) for other hypotheses and by Pyšek et al. (2008) and Bellard and Jeschke (2016) for the wider research field.

Our analysis did not show a decline in the level of empirical support over time. Jeschke et al. (2012a) reported such a decline for other invasion hypotheses, and decline effects have also been reported from a few other disciplines, particularly medicine, psychology and evolutionary ecology (Lehrer, 2010; Schooler, 2011). Underlying reasons include publication biases, biases in study organisms or systems and the psychology of researchers (Jeschke et al., 2012a, and references therein). Given that such a decline was not observed for the disturbance hypothesis suggests that our analysis has not provided an inflated level of empirical support for this hypothesis.

In a nutshell, the disturbance hypothesis is relatively well supported overall, at least when compared to other invasion hypotheses (cf. Jeschke et al., 2012a; and other chapters of this book). However, an empirical support by a bit more than half of the available studies cannot really be considered *strong* support. We encourage researchers to perform more experimental field studies in the future that focus on other types of disturbance than management or fire. Such studies will clarify the relationship between the type of methodological approach and level of support for the disturbance hypothesis.

References

Alston, K.P. and Richardson, D.M. (2006) The roles of habitat features, disturbance, and distance from putative source populations in structuring alien plant invasions at the urban/wildland interface on the Cape Peninsula, South Africa. *Biological Conservation* 132, 183–198.

Bellard, C. and Jeschke, J.M. (2016) A spatial mismatch between invader impacts and research publications. *Conservation Biology* 30, 230–232.

Burke, M.J.W. and Grime, J.P. (1996) An experimental study of plant community invasibility. *Ecology* 77, 776–790.

Dodge, R.S., Fulé, P.Z. and Sieg, C.H. (2008) Dalmatian toadflax (*Linaria dalmatica*) response to wildfire in a southwestern USA forest. *Ecoscience* 15, 213–222.

Elton, C.S. (1958) *The Ecology of Invasions by Animals and Plants*. Methuen, London.

Enders, M., Hütt, M.-T. and Jeschke, J.M. (2018) Drawing a map of invasion biology based on a network of hypotheses. *Ecosphere*. DOI: 10.1002/ecs2.2146.

Grime, J.P. (1977) Evidence for existence of three primary strategies in plants and its relevance to ecological and evolutionary theory. *American Naturalist* 111, 1169–1194.

Heger, T. and Jeschke, J.M. (2014) The enemy release hypothesis as a hierarchy of hypotheses. *Oikos* 123, 741–750.

Hernandez, R.R. and Sandquist, D.R. (2011) Disturbance of biological soil crust increases emergence of exotic vascular plants in California sage scrub. *Plant Ecology* 212, 1709–1721.

Hobbs, R.J. and Huenneke, L.F. (1992) Disturbance, diversity, and invasion: implications for conservation. *Conservation Biology* 6, 324–337.

Jeschke, J.M., Gómez Aparicio, L., Haider, S., Heger, T., Lortie, C.J., Pyšek, P. and Strayer, D.L. (2012a) Support for major hypotheses in invasion biology is uneven and declining. *NeoBiota* 14, 1–20.

Jeschke, J.M., Gómez Aparicio, L., Haider, S., Heger, T., Lortie, C. J., Pyšek, P. and Strayer, D.L. (2012b) Taxonomic bias and lack of cross-taxonomic studies in invasion biology. *Frontiers in Ecology and the Environment* 10, 349–350.

Johnson, P.T.J., Olden, J.D. and Vander Zanden, M.J. (2008) Dam invaders: impoundments facilitate biological invasions into freshwaters.

Frontiers in Ecology and the Environment 6, 357–363.

Lehrer, J. (2010) The truth wears off. *New Yorker*, Dec 13, 52–57.

Lowry, E., Rollinson, E.J., Laybourn, A.J., Scott, T.E., Aiello-Lammens, M.E., Gray, S.M., Mickley, J. and Gurevitch, J. (2013) Biological invasions: a field synopsis, systematic review, and database of the literature. *Ecology and Evolution* 3, 182–196.

Malumbres-Olarte, J., Barratt, B.I.P., Vink, C.J., Paterson, A.M., Cruickshank, R.H., Ferguson, C.M. and Barton, D.M. (2014) Big and aerial invaders: dominance of exotic spiders in burned New Zealand tussock grasslands. *Biological Invasions* 16, 2311–2322.

Moles, A.T., Flores-Moreno, H., Bonser, S.P., Warton, D.I., Helm, A., Warman, L., Eldridge, D.J., Jurado, E., Hemmings, F.A., Reich, P.B. *et al.* (2012) Invasions: the trail behind, the path ahead, and a test of a disturbing idea. *Journal of Ecology* 100, 116–127.

Pyšek, P., Richardson, D.M., Pergl, J., Jarošík, V., Sixtová, Z. and Weber, E. (2008) Geographical and taxonomic biases in invasion ecology. *Trends in Ecology & Evolution* 23, 237–244.

Rejmánek, M. (1989) Invasibility of plant communities. In: Drake, J.A., Mooney, H.A., di Castri, F., Groves, R.H., Kruger, F.J., Rejmánek, M. and Williamson, M. (eds) *Biological Invasions: a Global Perspective*. Wiley, Chichester, UK, pp. 369–388.

Schooler, J. (2011) Unpublished results hide the decline effect. *Nature* 470, 437.

Sher. A.A. and Hyatt. L.A. (1999) The disturbed resource-flux invasion matrix: a new framework for patterns of plant invasion. *Biological Invasions* 1, 107–114.

Shive, K.L., Kuenzi, A.M., Sieg, C.H. and Fulé, P.Z. (2013) Pre-fire fuel reduction treatments influence plant communities and exotic species 9 years after a large wildfire. *Applied Vegetation Science* 16, 457–469.

Sousa, W.P. (1984) The role of disturbance in natural communities. *Annual Review of Ecology and Systematics* 15, 353–391.

White, P.S. and Pickett, S.T.A. (1985) Natural disturbance and patch dynamics: an introduction. In: Pickett, S.T.A. and White, P.S. (eds) *The Ecology of Natural Disturbance and Patch Dynamics*. Academic Press, New York, pp. 3–13.

10 Invasional Meltdown Hypothesis

Raul Rennó Braga,[1,2]* **Lorena Gómez Aparicio,**[3]
Tina Heger,[4,5,6] **Jean Ricardo Simões Vitule,**[1,2] and
Jonathan M. Jeschke[7,8,6]

[1]*Programa de Pós-Graduação em Ecologia e Conservação,
Universidade Federal do Paraná, Curitiba, Brazil;* [2]*Laboratório
de Ecologia e Conservação, Depto de Engenharia Ambiental,
Setor de Tecnologia, Universidade Federal do Paraná,
Curitiba, Paraná, Brazil;* [3]*Department of Biogeochemistry,
Plant and Microbial Ecology, Institute of Natural Resources
and Agrobiology of Seville (IRNAS), CSIC, Sevilla, Spain;*
[4]*Biodiversity Research/Systematic Botany, University of
Potsdam, Germany;* [5]*Restoration Ecology, Technical Universiy
of Munich, Freising, Germany;* [6]*Berlin-Brandenburg Institute
of Advanced Biodiversity Research (BBIB), Berlin, Germany;*
[7]*Leibniz-Institute of Freshwater Ecology and Inland Fisheries
(IGB), Berlin, Germany;* [8]*Freie Universität Berlin, Institute of
Biology, Germany*

Abstract

Positive interactions among species can be central for community structure and ecosystem functioning. Given the current scenario of species invasions worldwide, the question arises how non-native species will interact in the new environment. Such reasoning has led to the invasional meltdown hypothesis (IM) which states that non-native species facilitate one another's invasion, increasing their likelihood of survival, ecological impact and possibly the magnitude of their impact. However, given the importance of antagonistic interactions in natural communities, it is not yet known to what extent these facilitative effects of non-native species occur. We used the hierarchy-of-hypotheses (HoH) approach to differentiate key aspects of IM and link empirical studies to specific sub-hypotheses of the overall hypothesis. Evidence related to IM was gathered by assessing citations of Simberloff and Von Holle (1999) who first defined it. Our HoH was categorized by the type of interaction among non-native species (e.g. facilitation, mutualism and multi-species interactions), ecological level of evidence and the outcome of the interaction for each non-native species (i.e. response variable measured in the study). We additionally looked for taxonomic and geographic variability. On the basis of the 208 relevant studies we found, the broad hypothesis and the majority of sub-hypotheses indicate that positive interactions among non-natives are happening more frequently than negative ones. Thus, IM is broadly supported by currently available studies. Evidence against IM relates to sub-hypotheses involving reciprocal interactions (e.g. competition between non-native species). We suggest that future research focuses on controlled experimental setups aiming at elucidating the community- or

* Corresponding author. E-mail: raulbraga@onda.com.br

ecosystem-level effects of non-native species interactions, especially for reciprocal interactions.

Introduction

Although species affect each other in both positive and negative ways, competition and predation have long been considered of major importance for community structure and organization. Their effects on population and community dynamics are well known. This rationale led to the formulation of the biotic resistance hypothesis (Elton, 1958; and Chapter 8, this volume). According to this hypothesis, as communities increase in species richness fewer available niches remain for arriving species and therefore establishment success is less probable (Fridley, 2011). However, more recently researchers have turned their attention to positive interactions such as mutualism and their importance for community processes (Bertness and Callaway, 1994; Bruno *et al.*, 2003). Facilitative interactions in the new environment can be crucial for the successful establishment of an arriving species. Thus, positive interactions can be central for community structure and ecosystem functioning (Christian, 2001; Hay *et al.*, 2004; Brooker *et al.*, 2008; Wright *et al.*, 2017).

Given the current acceleration of the movement of people and goods around the world and the associated increasing rate of invasions (Seebens *et al.*, 2017), an important question is how non-native species will interact in their new environment. For example, it is possible that the disruption of natural communities by already established non-native species will facilitate the arrival of newcomers. Furthermore, a cumulative effect is possible where many arriving non-native species will benefit from mutualistic interactions with previously established non-natives, resulting in a decreased number of failed introductions. The outcome of these facilitative interactions might be an increasing rate of non-native species able to establish in an area or a synergistic impact upon the native biota. This process has been called 'invasional meltdown' (Simberloff and Von Holle, 1999; Simberloff, 2006).

It has been stated that the invasional meltdown hypothesis (IM) is well supported by empirical evidence in comparison to other invasion hypotheses and that facilitative interactions among non-natives can be even more frequent than detrimental ones (Ricciardi, 2001; Jeschke *et al.*, 2012). In fact, invasional meltdown is considered to be one of the highest biodiversity threats in the UK for the next 50 years (Sutherland *et al.*, 2008). However, it remains unclear to what extent invasional meltdown is operating in invaded communities. For instance, Wonham and Pachepsky's (2006) null model indicates that an exponential trend in invasion records does not necessarily mean that there is invasional meltdown and, at least for animal interactions, invaders most commonly reduce one another's performance (Jackson, 2015), which is in contradiction to IM. Part of the contrasting results regarding this hypothesis might have arisen because the definition itself is very broad and open to different interpretations. The definition of invasional meltdown given by Simberloff and Von Holle (1999) and later updated by Simberloff (2006) is 'the process by which non-indigenous species facilitate one another's invasion in various ways, increasing the likelihood of survival and/or magnitude of impact and potentially leading to an accelerating increase in number of introduced species and their impact'.

Considering this definition, evidence can relate to three different aspects, and understanding their differences is crucial for the further development of the hypothesis. The first aspect relates to the type of the non-native species interactions reported, that is simple facilitation, mutual facilitation and a network of beneficial interactions. Although the original definition of the hypothesis says 'species facilitate one another', there are cases where only one-sided evidence is available (i.e. only evidence for how species A affects species B but not how species B affects species A). Studies providing this kind of evidence would only constitute weak evidence and should not be considered good examples for invasional meltdown (Simberloff, 2006), but none the less represent a scenario worth considering. Stronger evidence for IM comes from studies finding interactions with known effects for both species (i.e. A↔B). A deeper knowledge on the network of interactions within a community incorporating indirect effects among species would ultimately provide the strongest evidence in favour of IM.

Studies testing the species-interaction aspect of IM can further differ with respect to the ecological level they address. Researchers can search for empirical evidence evaluating non-native species interactions and their consequences for individual or population parameters. A species might for example increase one another's survival, growth, reproduction, abundance, biomass, density or other parameters.

The second aspect mentioned in the definition of IM is the *magnitude* of impact. This relates to the synergistic negative effect upon the native biota whereby the sum of the non-natives' individual impacts can be less than their effects when co-occurring (e.g. Johnson *et al.*, 2009; Jackson *et al.*, 2014).

The third and final aspect encapsulated in the definition of invasional meltdown is that the presence of non-natives can lead to an accelerated increase in non-native species richness (e.g. Ricciardi, 2001). Here, empirical evidence can arise through, for example, temporal trends in species richness. These different facets of the hypothesis can certainly be a major source of contradictory results.

In order to evaluate the usefulness of the hypothesis, it is helpful to separate its different aspects and so the hierarchy-of-hypotheses approach (see Chapters 2 and 6, this volume) was applied. To separate the main hypothesis into sub-hypotheses, we need to identify additional fundamental aspects that can contribute to the interpretation of patterns. The ecological level of evidence was already highlighted by Simberloff (2006) as being of importance for IM. An invasional meltdown would essentially be a community-level process, so evidence of non-natives affecting each other's survival or abundance, for example, would constitute

weaker evidence than the net effect of non-natives' interactions leading to increasing richness of non-native species. In this sense, studies providing evidence for invasional meltdown on community or ecosystem levels would constitute stronger cases of meltdowns. Our hierarchy of hypotheses (presented in the following), therefore, separates the main hypothesis into sub-hypotheses according to type of interaction, ecological level, and what was affected (i.e. response variable measured in the study) by the respective interaction. With this structure, we are able to evaluate the three main aspects of IM and gain insight into the mechanisms behind it.

The hierarchy of hypotheses presented here is an updated version of Braga *et al.* (2018) which generally forms the basis for this chapter. The current chapter includes information from 58 additional studies published after we completed the data collection for Braga *et al.* (2018). Geographic patterns are also featured here and taxonomic patterns are outlined in more detail than in Braga *et al.* (2018). Thus, the current chapter provides an up-to-date assessment of the empirical support and usefulness of the invasional meltdown hypothesis and its various sub-hypotheses.

Methods

To assess the empirical evidence for the hypothesis, we evaluated all published articles that cited Simberloff and Von Holle (1999). Because this publication is recognized as the first one using the term invasional meltdown to describe positive interactions among non-native species, most studies testing IM are likely to cite this paper. To search for these studies, we used the ISI Web of Science database. We did not include books, theoretical studies, meta-analyses or reviews in our analysis because such publications typically do not provide original empirical data. For a previous study (Braga *et al.*, 2018), we searched the Web of Science on 21 November 2014 which returned 637 papers citing Simberloff and

Von Holle (1999). Of those, we identified 150 relevant empirical studies that were included in Braga *et al.* (2018). We updated this dataset for the current chapter, searching the Web of Science on 30 September 2016, thereby finding 185 additional papers citing Simberloff and Von Holle (1999), thus 822 papers in total citing this paper. We found 58 relevant studies additional to Braga *et al.* (2018) that we included in the dataset, hence the updated dataset includes a total of 208 studies with relevant empirical information on the hypothesis. The dataset is freely available online at www.hiknowledge.org.

Following the approach outlined in Heger and Jeschke (2014), Braga *et al.* (2018) and Chapter 6 of this book, we classified the evidence reported in each study as either supporting (i.e. evidence is in line with the hypothesis), questioning (i.e. evidence is conflicting with the hypothesis), or being undecided (i.e. provided evidence both for and against the hypothesis).

The following three criteria were used to create the HoH (cf. Fig. 10.1):

1. Type of interaction:
   i. A→B, where two non-native species interact and only one is affected, with no evidence for the second species (e.g. simple facilitation).
   ii. A↔B, where two non-native species interact and both species are affected (e.g. mutualism).
   iii. Multi-species interaction, i.e. an interaction network between three or more non-native species (e.g. one species affects the interaction between the second and third species).
2. Ecological level that was affected:
   i. Individual.
   ii. Population.
   iii. Community.
   iv. Ecosystem.
3. Response variable measured in the study – this criterion relates to the effects of the interaction between the involved non-native species. We divided it according to ecological level (see 2):

Individual level:
   a. Resource (e.g. food source, feeding preference, predation, herbivory).
   b. Survival of individuals, growth or reproduction (e.g. interaction increases or decreases survival of non-natives).
   c. Dispersal of individuals (e.g. non-natives being dispersed by other non-natives but with no evaluated effect on population range expansion).
   d. Impact on individuals of native species (e.g. decreased or increased survival of natives was detected due to interacting non-natives).

Population level:
   a. Abundance, density or biomass (e.g. interaction increased or decreased abundance of non-natives).
   b. Population dispersal (e.g. non-natives being dispersed by other non-natives with effect on population range expansion).
   c. Impact on native population (e.g. increased or decreased abundance of native species was detected due to interacting non-natives).

Community level:
   a. Composition (i.e. the interaction of non-natives leads to compositional changes among the non-native species in the community).
   b. Richness (i.e. the interaction of non-natives increases or decreases the number of non-native species).
   c. Diversity (i.e. the interaction of non-natives changes the diversity of non-native species in another way).
   d. Impact on native community (e.g. the interaction of non-natives lead to compositional, richness or diversity changes of native species in the community).
   v. Ecosystem level: no further division was needed due to a lack of studies at this ecological level.

For eight papers, it was not possible to classify the outcome of the interaction, thus these were excluded from the analysis at this level (they were included for other analyses).

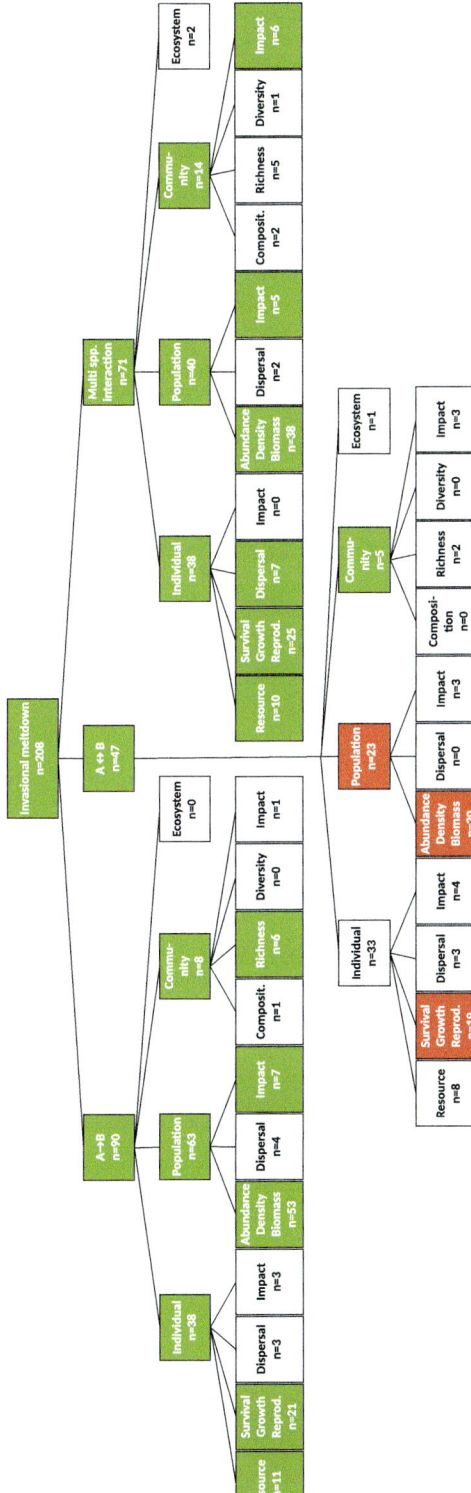

Fig. 10.1. Hierarchy of hypotheses for the invasional meltdown hypothesis (IM). The HoH is structured according to three criteria: (i) type of interaction (A→B, A↔B and multi-species interaction); (ii) ecological level of evidence (individual, population, community and ecosystem); and (iii) response variable measured in the study (resource, survival, growth, reproduction, dispersal, impact, abundance, density, biomass, composition, richness and diversity). Colour codes indicate levels of empirical support, as follows: green boxes, n≥5 and >50% of weighted evidence supporting the sub-hypothesis; red boxes, n≥5 and >50% of weighted evidence questioning the sub-hypothesis; white boxes, all other cases (all n<5, so no comparisons were made here).

With this HoH structure, we evaluated all main aspects encapsulated in the IM definition, namely the existence of facilitating species interactions (assessed by criterion 1), synergistic impact (termed 'magnitude of impact' by Simberloff and Von Holle, 1999; assessed by criterion 3 where it relates to impact on native individuals, population or community), and the accelerated increase in non-native species (assessed by criterion 3 where it relates to richness in the strict sense, and more broadly also to composition and diversity). As IM is, in Simberloff's (2006) words 'a community-level process', we evaluated if evidence so far has been gathered at this level (assessed by criterion 2).

Additional information was retrieved for each study concerning:

- Focal non-native species involved: taxonomic group (plants; algae; fungi; crustaceans, insects, molluscs, other invertebrates; fishes, amphibians, reptiles, birds, mammals; eubacteria/archaea/viruses), and number of species investigated.
- Study location: continents (North America, South America, Europe, Asia, Australia/Oceania, Africa, Antarctica) and major type of habitat (terrestrial, freshwater, marine).
- Research method: experiment or observation; conducted in the field, enclosure (incl. exclosure and common garden), or in the laboratory; and if the evidence provided was analysed quantitatively with statistics, quantitatively without statistics (e.g. due to small sample size) or only qualitatively (only non-numerical information is presented).

Using information on the research method, number of focal non-native species, type of interaction and ecological level, we weighted studies adapting the formula suggested by Heger and Jeschke (2014): study weight w was calculated as:

$$w = m \times \sqrt{n} \times i \times j \qquad (10.1)$$

where m is a score for the research method (1 for observational enclosure studies, 2 for observational field studies or experimental laboratory studies, 4 for experimental enclosure studies and 8 for experimental field studies), n is the number of focal non-native species involved (capped at a maximum value of 100), i is a score for the type of interaction (1 for A→B, 3 for A↔B and 8 for multi-species interactions) and j is a score for the ecological level (1 for individuals, 2 for populations, 6 for communities and 8 for ecosystems). For the research method and ecological level, studies sometimes presented information on more than one category and the highest value was used for weight calculations in these cases.

We chose to use the type of interaction for weight calculation because the clear distinction between A→B and A↔B has also been pointed out by Simberloff (2006) when stating that the first is a 'weaker version of meltdown'; multi-species interactions therefore constitute an even stronger version. Invasional meltdown is defined as a community-level phenomenon; weighting available evidence according to the ecological level therefore seems to be reasonable, with community-level studies receiving higher weights than studies at the population and individual level. If the outcome of the involved non-native species interaction leads to alterations at the ecosystem level, the community will also be affected; therefore evidence at this level is considered even stronger.

Weights varied from 3 to 1024 across all studies. To avoid an inflation of the sample size, we calculated proportional weights by dividing the separate sum of weights supporting, questioning or being undecided for a given sub-hypothesis by the total sum of weights of that sub-hypothesis. This result was multiplied by the sample-size number of the sub-hypothesis and rounded to integers (following Maletta, 2007; Heger and Jeschke, 2014).

Mann–Whitney U-tests were performed to test whether empirical support differs between sub-hypotheses. To assess whether results supporting, questioning or being undecided deviate from an equal distribution within each sub-hypothesis, we performed Chi (χ)-square tests. Post-hoc comparisons between supporting and questioning studies were carried out for statistically significantly differences.

Results

Out of the 208 relevant publications, 63.5% (n = 132) presented evidence supporting the hypothesis, 25.0% (n = 52) were questioning and 11.5% (n = 24) undecided. The results were similar when applying study weights, with 64.1% of studies supporting, 24.1% questioning IM and 11.8% being undecided (Table 10.1).

There was a clear geographic bias towards North America in the dataset, with almost half of the studies (46.6%, n = 97) conducted on this continent. Europe, in second place, had less than half of North America's studies with 17.8% (n = 37) (Fig. 10.2a). Field observational studies were the most common research method, representing 41.4% (n = 121) of the studies. Field experimental studies, the research method with the highest weight, was the second most common, representing 31.5% (n = 92) of the studies (Fig. 10.2b). The vast majority of studies presented evidence evaluated quantitatively with statistics (91.3%, n = 189).

For both unweighted and weighted data, A→B (n = 90) and multi-species interactions (n = 71) had significantly more studies supporting IM, whereas for A↔B (n = 47) there was no difference between the number of supporting vs questioning studies (Table 10.1). The level of support differed significantly between A→B and A↔B, with higher levels of support for one-sided interactions (Fig. 10.3a). Regarding ecological level, many studies presented evidence for more than one level. The majority of studies presented evidence at the population level (48.1%, n = 129) followed by individual (40.7%, n = 109), community (10.1%, n = 27) and, last, ecosystem level (1.1%, n = 3). All ecological levels had significantly more

Table 10.1. Weighted evidence from empirical tests supporting, questioning or being undecided about IM for each interaction type, ecological level and response variable measured in the study with χ^2 values for comparison of the distribution of the three categories to an equal distribution. χ^2 tests were only conducted for comparisons with more than five studies. Binomial tests comparing the proportions of supporting vs questioning studies were only conducted when χ^2 tests were significant ($p < 0.05$).

	n	Supporting	Undecided	Questioning	χ^2	Binomial test
Total	208	64.1%	11.8%	24.1%	<0.001	<0.001
A→B	90	72.0%	8.5%	19.5%	<0.001	<0.001
A↔B	47	52.6%	7.8%	39.6%	<0.001	0.365
Multi spp.	71	65.0%	12.7%	22.3%	<0.001	<0.001
Individual	109	61.0%	18.1%	20.8%	<0.001	<0.001
Resource	29	65.5%	29.4%	5.1%	<0.001	<0.001
Survival/growth/ reproduction	65	60.3%	16.3%	23.4%	<0.001	0.001
Dispersal	14	90.5%	7.6%	1.9%	<0.001	<0.001
Impact	8	50.4%	0.0%	49.6%	0.134	–
Population	129	69.2%	11.6%	19.2%	<0.001	<0.001
Abundance/density/ biomass	111	65.8%	13.7%	20.5%	<0.001	<0.001
Dispersal	6	96.2%	3.8%	0.0%	0.002	0.014
Impact	15	93.3%	0.2%	6.5%	<0.001	<0.001
Community	27	68.3%	6.3%	25.4%	<0.001	0.027
Composition	4	58.8%	41.2%	0.0%	–	–
Richness	13	56.7%	4.5%	38.9%	0.115	–
Diversity	1	100%	0.0%	0.0%	–	–
Impact	11	98.8%	0.0%	1.2%	<0.001	0.001
Ecosystem	3	100.0%	0.0%	0.0%	–	–

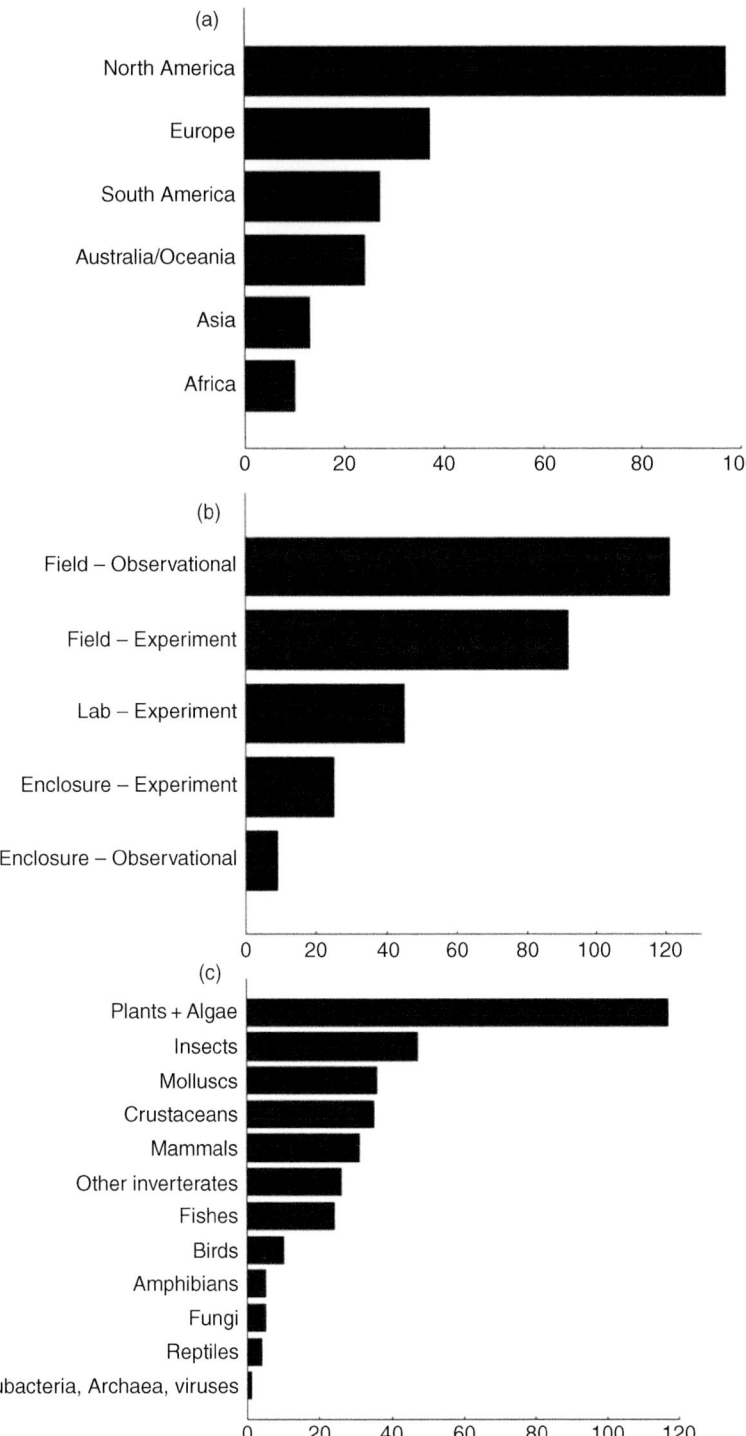

Fig. 10.2. Number of studies for the invasional meltdown hypothesis for (a) continents, (b) research methods and (c) taxonomic groups.

studies supporting than questioning IM, for both unweighted and weighted data. Additionally, the level of support did not change significantly among ecological levels (Fig. 10.3b).

After separating studies according to the major type of habitat, we saw a clear dominance of evidence for terrestrial (63.5%, n = 132) over freshwater (22.6%, n = 47) and marine habitats (13.9%, n = 29). Again, for all habitats, supporting studies were significantly more frequent than questioning ones. In these cases, however, the difference arose for freshwater and marine habitats only when considering weighted evidence (Table 10.2). Comparing the level of support among habitats, they were all similar (Fig. 10.3c).

Regarding taxonomic groups, plants and algae together were by far the most studied organisms when evaluating IM,

accounting for 34.3% (n = 117) of all studies. Insects (13.8%, n = 47), molluscs (10.6%, n = 36) and crustaceans (10.3%, n = 35) were the most studied invertebrates, whereas mammals with 9.1% (n = 31) were the most studied vertebrates (Fig. 10.2c). Among all the taxonomic groups evaluated, invertebrates (n = 117) got the highest level of support (Fig. 10.3d).

The HoH illustrates the number of studies and level of support for different sub-hypotheses (Fig. 10.1). Separating evidence by type of interaction, all sub-hypotheses with A→B and multi-species interactions, and with sufficient number of studies to be evaluated (n≥5), were supported by more than 50% of the empirical tests we found. However, A↔B interactions at population level (more precisely studies where two non-native species affect each other's abundance, density and/or biomass) and at the

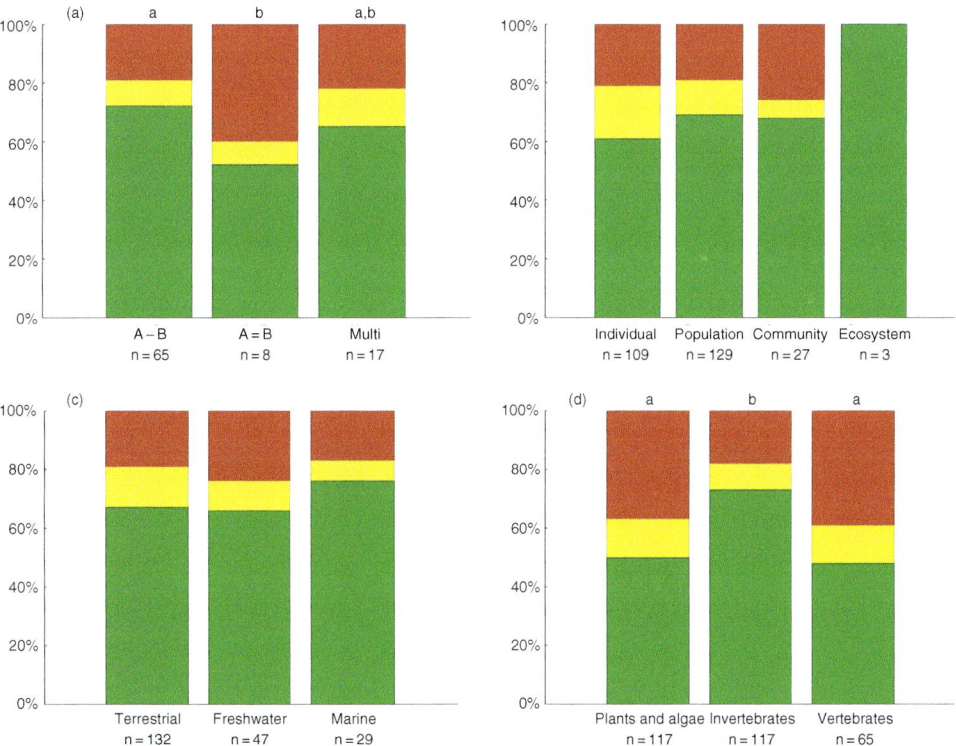

Fig. 10.3. Weighted data on empirical level of support for the invasional meltdown hypothesis, subdivided for (a) different types of interactions, (b) ecological level, (c) habitats and (d) taxonomic focus. Letters above bars indicate statistically significant differences (U-tests, p<0.05). Green represents supporting, yellow undecided and red questioning evidence.

Table 10.2. Weighted evidence from empirical tests supporting, questioning or being undecided about IM for habitats and taxonomic groups with χ^2 values for comparison of the distribution of the three categories to an equal distribution. χ^2 tests were only conducted for comparisons with more than five studies. Binomial tests comparing the proportion of supporting versus questioning studies were conducted when χ^2 tests were significant ($p < 0.05$).

	n	Supporting	Undecided	Questioning	χ^2	Binomial test
Terrestrial	132	67.2%	13.7%	19.0%	<0.001	<0.001
Freshwater	47	66.5%	9.8%	2.6%	<0.001	<0.001
Marine	29	75.5%	7.1%	17.4%	<0.001	0.001
Plants and algae	117	49.6%	12.8%	37.6%	<0.001	0.165
Invertebrates	117	73.5%	8.9%	17.6%	<0.001	<0.001
Vertebrates	65	48.6%	12.7%	38.7%	<0.001	0.353

individual level (studies showing two species affecting each other's survival, growth and/or reproduction) were less supported (i.e. >50% of empirical tests were questioning). The HoH also shows a low number of studies at community and ecosystem levels.

Discussion

Consistent with the number of Simberloff and Von Holle (1999) citations so far, a large body of evidence is available regarding facilitative non-native species interactions. Applying the HoH approach to the hypothesis, we were able to separate the main hypothesis into more specific sub-hypotheses that are more easily testable and comparable (see also Braga *et al.*, 2018). We were also able to identify gaps in research and to differentiate sub-hypotheses with varying degrees of empirical support.

The evidence available for the overarching hypothesis and the majority of sub-hypotheses indicates that IM is supported overall. These results are in accordance with Simberloff and Von Holle (1999), Ricciardi (2001), Jeschke *et al.* (2012) and Braga *et al.* (2018). The similarity in our findings to Braga *et al.* (2018) was expected because the current analysis is an update; however, some differences were found where previously the low number of studies limited the analyses. For example, four sub-hypotheses (richness at A→B, community at A↔B, population and community impact on multi-species) had more evidence (>50%)

supporting IM with the higher number of studies found in this updated analysis (see Fig. 10.1). Furthermore, significant differences not previously detected were now detected when comparing the level of support for survival, dispersal at population level, richness, freshwater and marine habitats, plants and algae, and vertebrates (Tables 10.1 and 10.2).

The large amount of evidence supporting IM for multi-species interactions strengthens the view that invasional meltdowns are widely happening within invaded communities. However, questioning evidence was dominant for sub-hypotheses with reciprocal interactions (A↔B). Indeed, this is the sub-hypothesis that incorporates a large amount of evidence for competing species and thus a dominance of negative interactions could be expected given its historical importance in explaining community structure (Elton, 1946; Diamond, 1975). Recent findings by Kuebbing and Nuñez (2016) suggest, however, that we should look at this result with caution. They found that competition between non-native species is less intense than competition between non-natives and native species. Thus, even though non-natives are negatively affecting each other, they can accumulate because the competition is more intense for native species. Such an accumulation of non-native species is not due to direct facilitation. Perhaps this scenario should be considered for a revised definition of IM. In addition, supporting evidence for this sub-hypothesis was dominated by plant–pollinator mutualisms, which have the potential to reach community alterations and thus constitute an invasional meltdown (e.g. Morales and Aizen, 2006). Given the results shown here, evidence for this sub-hypothesis (A↔B) should be further investigated even though it was not among the least studied sub-hypotheses.

Despite the observed pattern of support, there are still few studies on many sub-hypotheses. These gaps prevent further evaluation of the hypothesis and should be the focus of future studies. The shortage of studies at the community and ecosystem level is particularly crucial. The combined evidence of O'Dowd et al. (2003), Green et al. (2011) and O'Loughlin and Green (2015) are good examples for such studies. Indeed, information at these levels is harder to collect (Odum and Barret, 2005) but, as emphasized by Simberloff (2006), invasional meltdown is a community-level process and more studies at this level are most needed.

The geographic biases we found are largely in accordance with the biases found in the field of invasion biology in general, as well as in other research disciplines (SCImago, 2007; Pyšek et al., 2008; Lowry et al., 2013; Bellard and Jeschke, 2016). This was to be expected: the long history of ecological studies and financial resources in North America and Europe led to more studies in these continents, especially on such complex subjects as invasional meltdown. On the contrary, Brazil as an example of a country outside the North America–Europe axis hosts highly diverse environments but studies on non-native species are dominated by new records of invasions rather than complex interspecific relations within a community (Frehse et al., 2016). A relatively large number of studies on IM have been carried out in Oceania considering its small area. This is probably due to the large number of non-native species and their ecological consequences in this region (Kingsford et al., 2009).

Taxonomic and environmental biases arose owing to the large amount of studies related to interactions among terrestrial plants and insects. Although this is a general trend for invasion biology studies (e.g. Pyšek et al., 2008; Lowry et al., 2013; Frehse et al.,

2016), the bias found for IM was mainly due to pollination studies. Indeed, pollination interactions provide strong evidence for IM because generalist species such as bees are frequently able to form new mutualisms with non-native plants that have a generalist reproductive biology (Barthell et al., 2001; Olesen et al., 2002; Beavon and Kelly, 2012).

In conclusion, there is a large body of evidence demonstrating that invasional meltdown might play an important role in biological invasions across habitats and species groups. There are, however, several sub-hypotheses with inconclusive support owing to the low numbers of available studies – these should be of high priority for future studies. Of highest priority should be community-level studies because they constitute the core of IM. In addition, we suggest carrying out more controlled experiments aiming at elucidating the community- or ecosystem-level effects of non-native species interactions, especially for reciprocal interactions (A↔B).

References

Barthell, J.F., Randall, J.M., Thorp, R.W. and Wenner, A.M. (2001) Promotion of seed set in yellow star-thistle by honey bees: evidence of an invasive mutualism. *Ecological Applications* 11, 1870–1883.

Beavon, M.A. and Kelly, D. (2012) Invasional meltdown: pollination of the invasive liana *Passiflora tripartite* var. *mollissima* (Passifloraceae) in New Zealand. *New Zealand Journal of Ecology* 36, 100–107.

Bellard, C. and Jeschke, J.M. (2016) A spatial mismatch between invader impacts and research publications. *Conservation Biology* 30, 230–232.

Bertness, M.D. and Callaway, R. (1994) Positive interactions in communities. *Trends in Ecology & Evolution* 9, 191–193.

Braga, R.R., Gómez-Aparicio, L., Heger, T., Vitule, J.R.S. and Jeschke, J.M. (2018) Structuring evidence for invasional meltdown: broad support but with biases and gaps. *Biological Invasions.* DOI:10.1007/s10530-017-1582-2

Brooker, R.W., Maestre, F.T., Callaway, R.M., Lortie, C.L., Cavieres, L.A. *et al.* (2008) Facilitation in plant communities: the past, the present, and the future. *Journal of Ecology* 96, 18–34.

Bruno, J.F., Stachowicz, J.J. and Bertness, M.D. (2003) Inclusion of facilitation into ecological theory. *Trends in Ecology & Evolution* 18, 119–125.

Christian, C.E. (2001) Consequences of a biological invasion reveal the importance of mutualism for plant communities. *Nature* 413, 635–639.

Diamond, J.M. (1975) Assembly of species communities. In: Cody, M.L. and Diamond, J.M. (eds) *Ecology and Evolution of Communities*. Harvard University Press, Cambridge, Massachusetts, pp. 342–444.

Elton, C. (1946) Competition and the structure of ecological communities. *Journal of Animal Ecology* 15, 54–68.

Elton, C.S. (1958) *The Ecology of Invasions by Animals and Plants*. Methuen, London.

Frehse, F.A., Braga, R.R., Nocera, G.A. and Vitule, J.R.S. (2016) Non-native species and invasion biology in a megadiverse country: scientometric analysis and ecological interactions in Brazil. *Biological Invasions* 18, 3713–3725.

Fridley, J.D. (2011) Invasibility, of communities and ecosystems. In: Simberloff, D. and Rejmánek, M. (eds) *Encyclopedia of Biological Invasions*. University of California Press, Los Angeles, California, pp. 356–360.

Green, P.T., O'Dowd, D.J., Abbott, K.L., Jeffery, M., Retallick, K. *et al.* (2011) Invasional meltdown: invader-invader mutualism facilitates a secondary invasion. *Ecology* 92, 1758–1768.

Hay, M.E., Parker, J.D., Burkepile, D.E., Caudill, C.C., Wilson, A.E. *et al.* (2004) Mutualisms and aquatic community structure: the enemy of my enemy is my friend. *Annual Review of Ecology, Evolution, and Systematics* 35, 175–197.

Heger, T. and Jeschke J.M. (2014) The enemy release hypothesis as a hierarchy of hypotheses. *Oikos* 123, 741–750.

Jackson, M.C. (2015) Interactions among multiple invasive animals. *Ecology* 96, 2035–2041.

Jackson, M.C., Jones, T., Milligan, M., Sheath, D., Taylor, J., Ellis, A., England, J. and Grey, J. (2014) Niche differentiation among invasive crayfish and their impacts on ecosystem structure and functioning. *Freshwater Biology* 59, 1123–1135.

Jeschke, J.M., Gómez Aparicio, L., Haider, S., Heger, T., Lortie, C.J., Pyšek, P. and Strayer, D.L. (2012) Support for major hypotheses in invasion biology is uneven and declining. *NeoBiota* 14, 1–20.

Johnson, P.T.J., Olden, J.D., Solomon, C.T. and Vander Zanden, M.J. (2009) Interactions among invaders: community and ecosystem effects of multiple invasive species in an experimental aquatic system. *Oecologia* 159, 161–170.

Kingsford, R.T., Watson, J.E.M., Lundquist, C.J., Venter, O., Hughes, L., Johnston, E.L., Atherton, J., Gawel, M., Keith, D.A., Mackey, D.G. *et al.* (2009) Major conservation policy issues for biodiversity in Oceania. *Conservation Biology* 23, 834–840.

Kuebbing, S.E. and Nuñez, M.A. (2016) Invasive non-native plants have a greater effect on neighbouring natives than other non-natives. *Nature Plant,* 2, 16134.

Lowry, E., Rollinson, E.J., Laybourn, A.J., Scott, T.E., Aiello-Lammens, M.E., Gray, S.M., Mickley, J. and Gurevitch, J. (2013) Biological invasions: a field synopsis, systematic review, and database of the literature. *Ecology and Evolution* 3, 182–196.

Maletta, H. (2007) Weighting. Available at: www.spsstools.net/Tutorials/WEIGHTING.pdf (accessed 5 October 2017).

Morales, C.L. and Aizen, M.A. (2006) Invasive mutualisms and the structure of plant-pollinator interactions in the temperate forests of northwest Patagonia, Argentina. *Journal of Ecology* 94, 171–180.

O'Dowd, D.J. and Green, P.T. and Lake, P.S. (2003) Invasional 'meltdown' on an oceanic island. *Ecology Letters* 6, 812–817.

Odum, E.P. and Barrett, G.W. (2005) *Fundamentals of Ecology*. Thomson Brooks, Cole, California.

Olesen, J.M., Eskildsen, L.I. and Venkatasamy. S. (2002) Invasion of pollination networks on oceanic islands: importance of invader complexes and endemic super generalists. *Diversity and Distributions* 8, 181–192.

O'Loughlin, L.S. and Green, P.T. (2015) Invader-invader mutualism influences land snail community composition and alters invasion success of alien species in tropical rainforest. *Biological Invasions* 17, 2659–2674.

Pyšek, P., Richardson, D.M., Pergl, J., Jarošík, V., Sixtová, Z. and Weber, E. (2008) Geographical and taxonomic biases in invasion ecology. *Trends in Ecology & Evolution* 23, 237–244.

Ricciardi, A. (2001) Facilitative interactions among aquatic invaders: is an "invasional meltdown" occurring in the Great Lakes? *Canadian Journal of Fisheries and Aquatic Sciences* 58, 2513–2525.

SCImago (2007) SJR — SCImago Journal & Country Rank. Available at: http://www.scimagojr.com (accessed 5 October 2017).

Seebens, H., Blackburn, T.M., Dyer, E.E., Genovesi, P., Hulme, P.E., Jeschke, J.M., Pagad, S., Pyšek, P., Winter, M., Arianoutsou, M. *et al.* (2017) No saturation in the accumulation of alien species worldwide. *Nature Communications* 8, 14435.

Simberloff, D. (2006) Invasional meltdown 6 years later: important phenomenon, unfortunate metaphor, or both? *Ecology Letters* 9, 912–919.

Simberloff, D. and Von Holle, B. (1999) Positive interactions of nonindigenous species: invasional meltdown? *Biological Invasions* 1, 21–32.

Sutherland, W.J., Bailey, M.J., Bainbridge, I.P., Brereton, T., Dick, J.T.A., Drewitt, J., Dulvy, N.K., Dusic, N.R., Freckleton, R.P., Gaston, K.J. *et al.* (2008) Future novel threats and opportunities facing UK biodiversity identified by horizon scanning. *Journal of Applied Ecology* 45, 821–833.

Wonham, M. J. and Pachepsky, E. (2006) A null model of temporal trends in biological invasion records. *Ecology Letters* 9, 663–672.

Wright, A.J., Wardle, D.A., Callaway, R. and Gaxiola, A. (2017) The overlooked role of facilitation on biodiversity experiments. *Trends in Ecology & Evolution* 32, 383–390.

11

Enemy Release Hypothesis

Tina Heger[1,2,3]* and Jonathan M. Jeschke[4,5,3]

[1]University of Potsdam, Biodiversity Research/Systematic Botany, Potsdam, Germany; [2]Technical University of Munich, Restoration Ecology, Freising, Germany; [3]Berlin-Brandenburg Institute of Advanced Biodiversity Research (BBIB), Berlin, Germany; [4]Leibniz-Institute of Freshwater Ecology and Inland Fisheries (IGB), Berlin, Germany; [5]Freie Universität Berlin, Institute of Biology, Berlin, Germany

Abstract

The enemy release hypothesis is a prominent explanation for invasion success. It is, however, a complex multi-faceted hypothesis and has multiple sub-hypotheses. Empirical tests of the enemy release hypothesis therefore often address very different questions. For this reason, we previously applied the hierarchy-of-hypotheses (HoH) approach to analyse the level of evidence for the enemy release hypothesis and its sub-hypotheses, taking into account this variety of formulations and research questions. This chapter provides an update and extension of that analysis by including recently published evidence and implementing some of the suggestions made in Chapters 3 to 6, this volume, concerning the HoH approach. In detail, we: (i) re-organized the HoH, now separating research approaches from working hypotheses; (ii) displayed results including bar charts in order to avoid a classification of the level of evidence for the different hypotheses; and (iii) tested the robustness of our results based on the precision of the tests (estimated as number of replicates), their generality (using number of focal species as a proxy), realism (i.e. whether they were conducted in the lab, an enclosure or in the field) and their general approach (observation or experiment). We found relatively strong support for the enemy release hypothesis in studies looking at release in the sense of reduced enemy pressure. However, this is not paralleled by strong evidence for enhanced performance. Support for the enemy release hypothesis differs according to which question is asked. It is highest if species in their native vs introduced range are compared and lowest if invasive aliens are compared to non-invasive aliens. There is a comparatively high level of supporting evidence if specialist enemies are considered. From these findings, we conclude that future studies should focus on testing whether invaders show enhanced performance if released from specialist enemies – an underexplored research question so far. Our robustness analysis indicates that empirical results are influenced by the generality (estimated by the number of focal species) and realism (estimated by the number of replicates) of studies addressing the enemy release hypothesis. We suggest that future studies in the context of enemy release should preferentially be done in the field rather than in the lab and use more than ten focal alien species.

Introduction

In broad terms, the enemy release hypothesis posits that the absence of enemies is a

---

* Corresponding author. E-mail: tina-heger@web.de

cause of invasion success (Keane and Crawley, 2002; Mitchell and Power, 2003; Torchin *et al.*, 2003; Colautti *et al.*, 2004; Liu and Stiling, 2006; Jeschke *et al.*, 2012; Heger and Jeschke, 2014; Jeschke, 2014). The underlying idea of this hypothesis is that alien species should be released from enemy pressure in the new range because the enemies of the species in their native range are not usually transported across the dispersal barrier together with the alien species and because the enemies in the new range should lack adaptations to the novel species and thus cannot readily use it as food or as a host species ('enemies' can be predators, herbivores, parasites, pathogens or parasitoids). The resulting reduction of enemies might lead to an advantage for the invader in its new range.

In 1915, the Swiss botanist Albert Thellung wrote that 'the success of alien species may rely on their lacking competitors or enemies in the new range' (p. 1249 in Kowarik and Pyšek, 2012). The enemy release hypothesis has, however, only been formalized as a general hypothesis in the 21st century (e.g. Keane and Crawley, 2002). In its modern formulation, it usually considers the *interaction* of alien species with native species in their new range (cf. Jeschke *et al.*, 2012).

The hypothesis is very important to the field of invasion biology today. In a recent survey, enemy release was indicated by 150 of 357 invasion biologists (42%) as one of up to three invasion hypotheses they know best (each survey participant could select up to three out of 33 presented invasion hypotheses). No other hypothesis received such a high score in the survey (propagule pressure, see Chapter 16, this volume, was second with 38% and the disturbance hypothesis, see Chapter 9, was third with 29%; Enders *et al.*, 2018). In Lowry *et al.*'s (2013) systematic review, enemy release ranked as fifth compared to other invasion hypotheses. This review covered all time periods and was not restricted to the most recent literature. Because the enemy release hypothesis has become a key invasion hypothesis only relatively recently, this systematic review seems to underestimate the

role that this hypothesis plays in the field today. It currently seems to be among the top two or three of the most important and popular hypotheses in invasion biology.

Because the general idea underlying the enemy release hypothesis has many implications and facets, there are many ways in which it can be addressed with empirical studies. In previous studies, we made use of the hierarchy-of-hypotheses (HoH) approach (Chapter 2, this volume; Jeschke *et al.*, 2012; Heger *et al.*, 2013a) to assess how much supporting evidence there is for the enemy release hypothesis, taking into account differences in empirical approaches (Jeschke *et al.*, 2012; Heger and Jeschke, 2014). In this chapter, we will further update and thus extend the dataset and, more importantly, will implement some of the suggestions on how to improve and refine the HoH approach outlined in Chapters 3 to 6, this volume.

Methods: Updating the HoH for Enemy Release

To update the previous dataset used for Heger and Jeschke (2014), we performed a search in the Web of Science using the same search terms as previously employed ('enemy release AND (alien OR exotic OR introduc* OR invas* OR naturali?ed OR nonindigenous OR non-indigenous OR nonnative OR non-native)'), but updating it to 2013 to 2016 (date of search: 15 November 2016). The search returned 398 hits. We checked for all of these references whether they addressed the enemy release hypothesis according to its broad formulation given above ('the absence of enemies is a cause of invasion success'). Some studies widened the hypothesis to include feeding preferences of resident species (e.g. food choice experiments), susceptibility of aliens to diseases, or resistance or tolerance of species to herbivory. We did not include such studies in our dataset. Soil-feedback studies were only included in cases where the authors were able to differentiate between negative (enemy) and positive effects of soil biota.

Also, studies on range expansions were excluded. In total, we identified 52 new relevant studies that we added to our dataset, which now included 163 studies. The dataset is freely available online at www.hi-knowledge.org.

As in Jeschke *et al.* (2012) and Heger and Jeschke (2014), each empirical test in a study was included as a separate entry in the dataset, yielding a total of 248 empirical tests. If a study reported multiple tests for one working hypothesis using the same research approach (Fig. 11.1), data for these tests were combined. To avoid inflation of sample sizes, we combined multiple tests from single studies at the higher levels of the HoH. For example, if a study reported results from a test focusing on release from specialist enemies and from a second test on release from generalist enemies, data from these two tests were combined for assessing the higher-level working hypothesis 'invaders are released from enemies'. Such a correction of sample sizes was not done in Jeschke *et al.* (2012) or Heger and Jeschke (2014).

For each test, we recorded whether the results are supporting or questioning the enemy release hypothesis, or whether evidence is undecided (e.g. because there was lower infestation with one parasite but higher infestation with a second parasite, as e.g. in Clark *et al.*, 2015). As outlined in Chapter 6, this volume, this approach differs from vote counting, which is only based on significance values and has key weaknesses. The approach applied here takes all available evidence into account, particularly effect sizes, to classify studies as supporting, questioning or being undecided.

We changed the basic structure of the HoH shown in Heger and Jeschke (2014) in accordance with the suggestions made in Chapters 3, 4 and 6 (Fig. 11.1). Measured response variables (damage by or infestation with enemies) and the type of comparison (e.g. alien vs native species or alien species in their native vs invaded range) are now no longer used as branching criteria creating separate sub-hypotheses. Instead, they are considered as parallel research approaches, and we assessed levels of

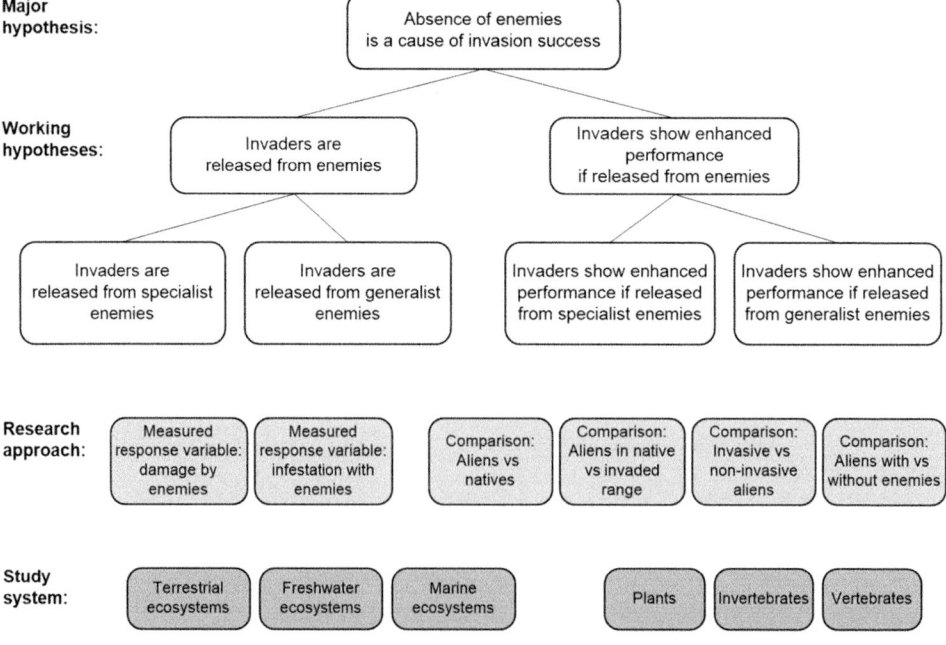

Fig. 11.1. The enemy release hypothesis as an HoH. This updated version (compared to Heger and Jeschke, 2014) implements suggestions from Chapters 3, 4 and 6, this volume.

empirical evidence for the major and the working hypotheses for these research approaches separately. As in Heger and Jeschke (2014), we also examined evidence for the major hypothesis in each taxonomic group and habitat type separately, and compared the results for these groups using pairwise U-tests (function 'wilcox.test' in R 3.1.2; R Core Team, 2014).

We had the additional aim to assess whether the level of empirical evidence received from the studies depends on the level of precision, realism and generality of the study, or on whether an observation or an experiment was done (see Chapters 4 and 6 for an explanation of the underlying ideas). For each empirical test, we therefore recorded the number of replicates (precision), the number of focal species (generality), whether the study was done in the lab, enclosures (e.g. common garden) or in the field, and whether it was an observational study or an experiment. Studies that compiled literature data (e.g. from databases as in Mitchell and Power, 2003) received a precision score of zero; observational studies done in one area a precision score of one. These analyses replace the weighting system we applied in Heger and Jeschke (2014), and thus avoid problems with this method pointed out in Chapters 4 and 5. With Kruskal–Wallis tests (function 'kruskal.test') and subsequent pairwise comparisons with Tukey and Kramer (Nemenyi) tests (function 'posthoc.kruskal.nemenyi.test', package 'PMCMR'; Pohlert, 2014), we checked whether the level of support for the enemy release hypothesis depends on where (lab, enclosure or field) and how (experiment or observation) an empirical test was done.

Results: Mixed Support for the Enemy Release Hypothesis

The level of support for the enemy release hypothesis and the refined working hypotheses derived from the assessment of the published empirical tests is shown in Figs 11.2 and 11.3, in which we chose to show the respective percentages of studies supporting, being undecided and not supporting the respective hypothesis in bar charts. This visualization of the results differs from Heger and Jeschke (2014) and from other chapters of this book where hypotheses are assigned to one of the three categories (supported, undecided or questioned) and thus displayed in either green, white or red, once the level of evidence has reached 50% for this category. The first graph (Fig. 11.2a) shows the level of evidence for the overarching enemy release hypothesis and the working hypotheses across all research approaches. The other graphs show results for subsets of the data, depending on which response variable has been measured (damage or infestation, Figs 11.2b,c) or which comparison has been made (Fig. 11.3). There were only five studies comparing the performance of aliens to those without enemies; we therefore did not produce a separate figure for this comparison. For working hypotheses not given in the graph, there were no empirical tests available.

Overall, 40% of the analysed empirical tests provided supporting evidence for the enemy release hypothesis (Fig. 11.2a). Compared to this result for the overall hypothesis, there was more support (48%) for the working hypothesis that invaders are released from enemies, and less support for the working hypothesis that invaders show enhanced performance if released from enemies (24%; 54% questioning evidence). The percentage of tests delivering supporting evidence was highest for the hypothesis that invaders are released from specialist enemies (50%). Release from generalist enemies has not been found in any test; however, sample size was rather low here ($n = 6$). In contrast, there was some evidence supporting the hypothesis that invaders show enhanced performance if released from generalist enemies (33%, $n = 6$). For both response variables that had been used to estimate release from enemies, nearly the same level of support was found (48% supporting for tests analysing damage, Fig. 11.2b, and 52% for those analysing infestation, Fig. 11.2c).

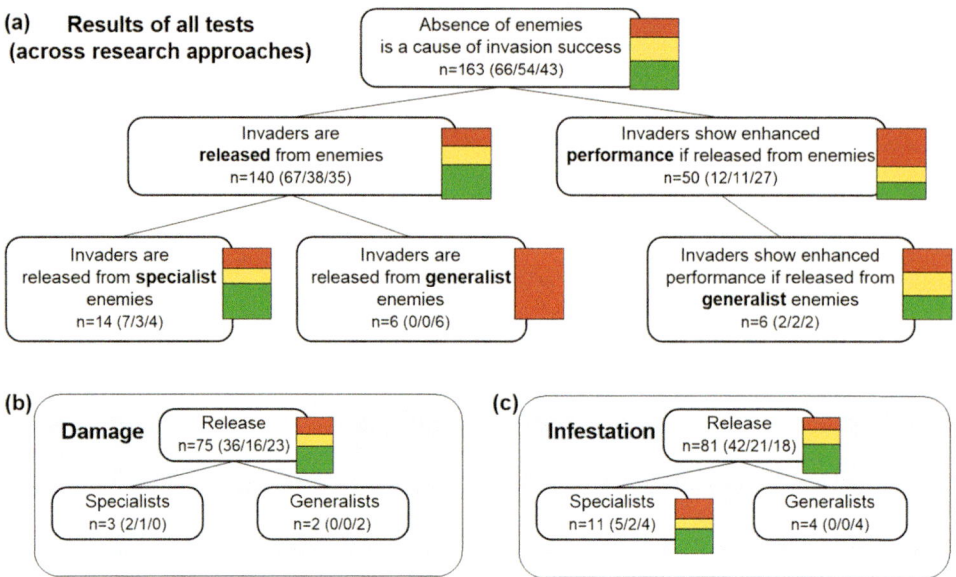

Fig. 11.2. Synthesized results of all analysed studies testing the enemy release hypothesis (a). For the working hypothesis 'Invaders are released by enemies', results are additionally shown separately depending on whether (b) damage of the invader (e.g. lost leaf tissue) has been measured or (c) infestation, i.e. occurrence of enemies on the invader. The bar charts next to each hypothesis show the percentages of studies supporting (green), being undecided (yellow) or questioning (red). Bar charts are not shown if n<5. In each box, the number of studies underlying the bar charts is given, as well as the number of tests supporting/being undecided/questioning the hypothesis (in brackets).

The highest level of support was found for studies comparing aliens in the native vs invaded range in terms of release (less damage or infestation; 60% supporting evidence; Fig. 11.3b). The lowest level of support was found for studies comparing invasive with non-invasive alien species (80% questioning Fig. 11.3c; but sample size only n = 15), and for those comparing performance of aliens vs natives (72% questioning evidence; Fig. 11.3a). In all comparisons, the same overall pattern was visible (cf. Fig. 11.2): there was more support for the hypothesis that invaders are less damaged by or infested with enemies than there was for the hypothesis that invaders show enhanced performance in response to lower enemy pressure. Release from specialist enemies was found even more frequently for all comparisons except for invasive vs non-invasive aliens: invasive aliens were not more frequently released from specialist enemies than non-invasive aliens.

Regarding study systems, the highest frequency of supporting studies for the overarching enemy release hypothesis was found for vertebrates (Fig. 11.4a); however, there was no statistically significant difference among taxonomic groups or habitats (Fig. 11.4a,b). This result differs from Lafferty *et al.* (2010), who found significantly more release (termed 'escape' in this publication) in aquatic than in terrestrial habitats. The enemy release hypothesis has been tested most frequently for plants and in terrestrial habitats.

Do results differ depending on precision, realism and generality of the studies, or on whether they are experiments or observations?

The precision of an empirical test (here estimated as number of replicates) does not seem to influence whether a study produced

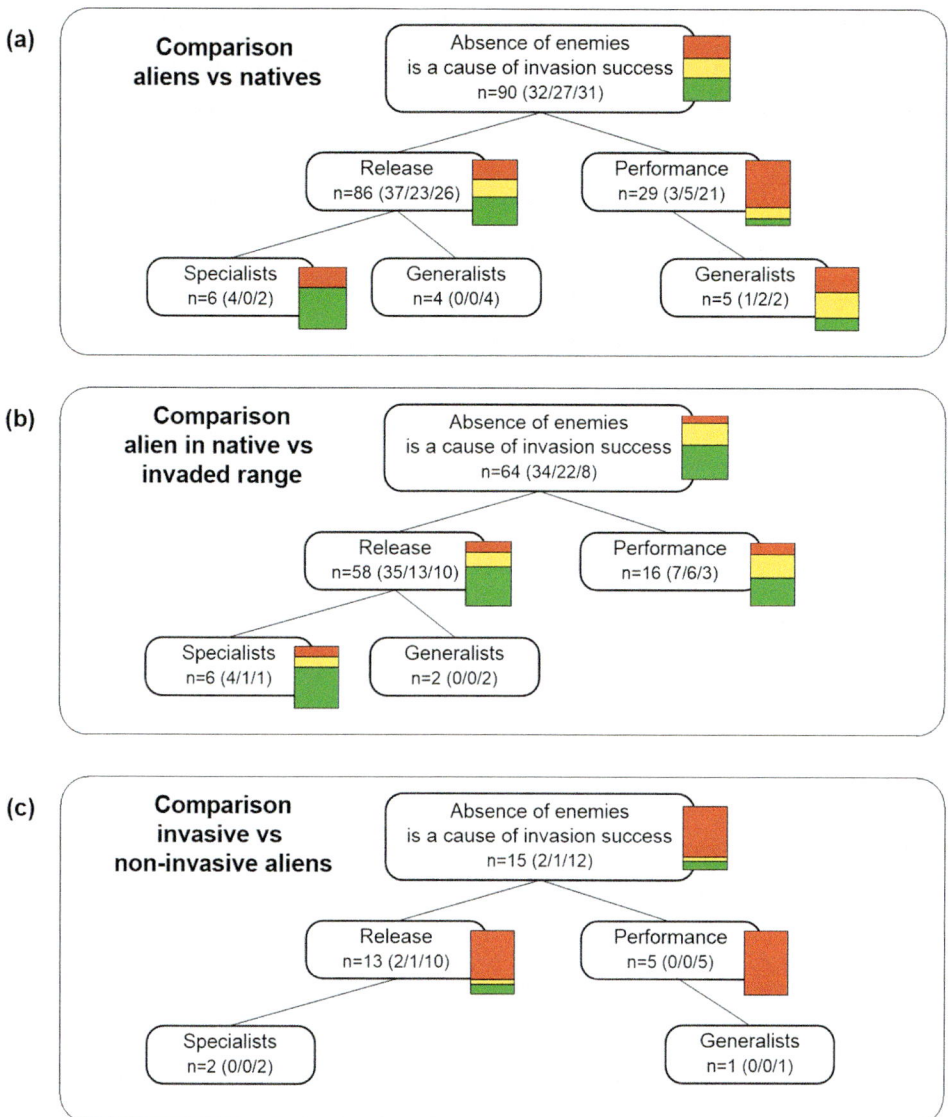

Fig. 11.3. Results of studies testing the enemy release hypothesis by comparing (a) alien with native species, (b) alien species in their native vs their invaded range and (c) invasive vs non-invasive alien species. For an explanation of the bar charts and numbers see Fig. 11.2.

support, no support or inconclusive results (Fig. 11.5a). More general studies (i.e. those focusing on multiple species within one study) produce more supporting evidence than studies focusing on one or a few focal species (Fig. 11.5b). It has to be noted, though, that the number of focal species does not only relate to the generality of a study but also to its realism. The more species are studied, the more likely the test is to detect species interactions that influence invasion success or failure.

Supporting evidence tends to stem from field studies (here interpreted as most realistic), whereas studies in enclosures tend to yield more questioning evidence. In the

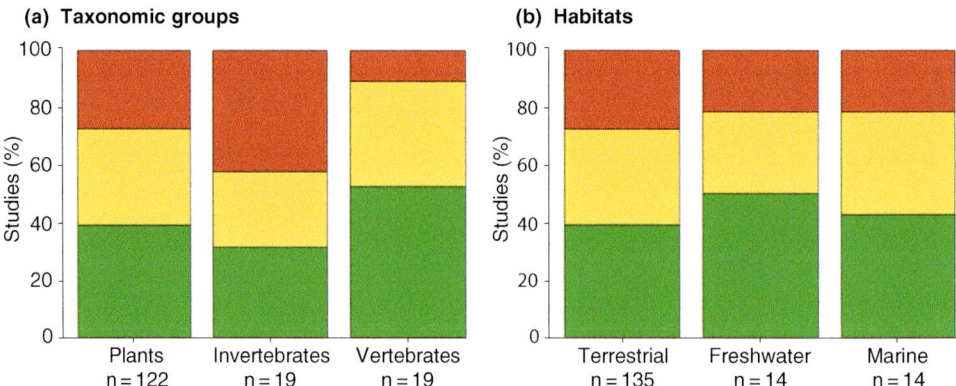

Fig. 11.4. Results of studies testing the enemy release hypothesis, shown separately for (a) taxonomic groups and (b) habitats. Pairwise U-tests did not show significant differences.

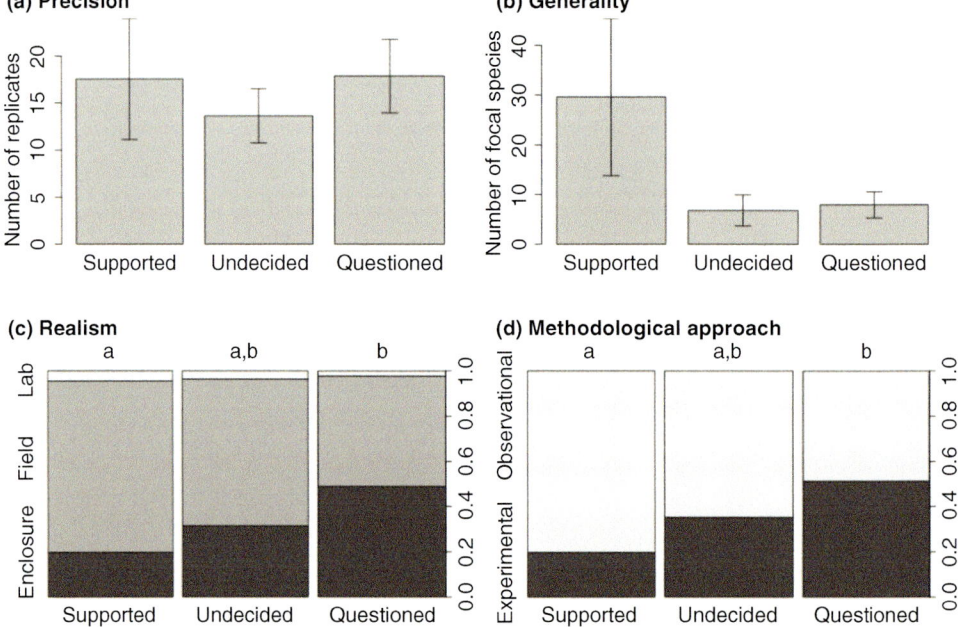

Fig. 11.5. Effects of (a) precision, (b) generality, (c) realism and (d) methodological approach (experiment or observation) on the level of support for the overall enemy release hypothesis; (a) and (b) show means and 95% confidence intervals, (c) and (d) show percentages (see y-axes on the right). Letters indicate significant differences ($p < 0.05$) in Tukey and Kramer (Nemenyi) tests.

analysed studies, realism is closely coupled to the methodological approach (experimental or observational), which is common in ecology. Experiments are usually done in labs or enclosures, whereas observations are typically done under field conditions. Thus, observations (in the field) in the analysed studies tended to yield supporting evidence, whereas experiments tended to produce questioning evidence. We found very similar patterns if the two working hypotheses ('release' and 'performance') were analysed separately (data not shown).

Discussion

Overall, our results showed different levels of support depending on which aspect of the enemy release hypothesis was tested, how it was tested and which species were examined. Compared to our previous study (Heger and Jeschke, 2014), the level of support was slightly higher (now 40% supporting evidence compared to previously 36%), and the percentage of tests questioning this major hypothesis was lower (now 26%, previously 43%). These changes in proportions are mainly caused by the methodological change described above: for the current analyses, we combined multiple tests published in one study at the higher levels of the hierarchy, to avoid inflation of sample sizes. In the previous study, the level of support for the major hypothesis was derived from 176 empirical tests. The combination of multiple tests on the higher levels of the hierarchy led to a decrease of this number to a total of 111 tests (not including the references added after the recent search in Web of Science). Because many studies published tests differing in their outcome, the combination of multiple tests also changed the percentages of the level of evidence. Re-analysing these 111 tests of the old dataset, we found 42% of the tests supporting the enemy release hypothesis, 35% being undecided, and 23% questioning it. These values are close to the results presented above. At the same time, they show that empirical support for the enemy release hypothesis tends to further decline over time: Jeschke et al. (2012) already reported a decline in empirical support. We can conclude that this decline has further continued. Decline effects have been observed in many disciplines and are usually attributed to a publication bias. When a hypothesis appears in the literature, studies supporting it can be published fast and easily, whereas after some time, results questioning it are attractive and thus easy to publish (see Jeschke et al., 2012).

Consistent with the patterns reported in Heger and Jeschke (2014), we found that overall, invaders are less damaged or infested by enemies, but this seems not to be linked to an enhanced performance. In addition, tests comparing aliens in their native and invaded range yield more supporting evidence than other comparisons (this has also been shown by Colautti et al., 2004). With increased sample size, it has now become clear that support for enemy release of invasive compared to alien but non-invasive species is generally weak and that evidence for release (in terms of damage and infestation) from specialists is comparatively strong.

The low level of support for a lower performance as a result of enemy release might indicate that performance is influenced by a variety of different factors, thus making tests of what affects performance very difficult. It is striking that results furthermore differ depending on which groups of species are compared. The high level of supporting evidence in studies comparing species in their native and invaded range indicate that in many cases invaders do leave behind their enemies. If compared to native species, however, the difference is less clear. An explanation could be that there is a flip side to the coin: not only is the invader novel for the resident predators, potentially causing release, but also the resident predators are novel for the invader (Saul et al., 2013); invaders might therefore lack eco-evolutionary experience and suffer relatively more from resident predators that are able to prey on them than native species (Colautti et al., 2004 and references therein). Furthermore, the comparison of different species within one range may be constrained by differences among these species in their life history, morphology or other traits. Many studies try to account for this problem by comparing closely related species with each other; still, this comparison might yield less reliable results than the comparison of species between their native and invaded range.

Besides the potential problem of comparing different species with each other, the interpretation of studies that address enemy release by comparing invasive with non-invasive aliens is further complicated by the fact that, across studies, different definitions for 'invasive' have been used. In many studies, species are defined as 'invasive' only if they have some type of negative ecological impact (e.g. Cappuccino and Carpenter,

2005). In other studies, however, species are classified as 'invasive' as soon as they are common and spreading (e.g. Liu *et al.*, 2006; see also Heger *et al.*, 2013b). Having these limitations in mind, we can cautiously conclude that our results do not suggest that enemy release contributes to a stronger negative ecological impact of alien species or to their enhanced spread within the invaded area (see also Merella *et al.*, 2016).

The result that release (in the sense of enemy reduction) is found more frequently than enhanced performance of released invaders indicates that the two working hypotheses of the overarching enemy release hypothesis – i.e. 'invaders are released from enemies' and 'invaders show enhanced performance if released from enemies' – are actually not as tightly linked as suggested by the enemy release hypothesis. Enemy release is multifaceted and complex, and support for one aspect of the general hypothesis does not necessarily translate to another aspect. To assess the idea underlying the enemy release hypothesis, it therefore seems fruitless to just focus on the first working hypothesis that invaders are released from enemies; instead, we recommend directly focusing on the performance of invaders and how this is affected by enemies. Because lower infestation and damage has been found particularly frequently for specialist enemies, it would be interesting to test whether a release from specialist enemies leads to an enhanced performance of invaders. We found no study testing this specific hypothesis.

According to our results, studies more often yield supporting results if they are using multiple species; undecided and questioning results were more often reported by studies with fewer than ten focal species. This is an interesting result, potentially indicating some bias in the choice of focal species. Looking at observational and experimental tests separately, this relationship is found only for observational studies where the number of focal species was generally higher (21.6 species on average compared to 5.8 for experimental studies). The fact that field studies and observations yielded more evidence supporting the enemy release

hypothesis could be due to a strengthening of enemy release by some interacting factors in the field, e.g. resource availability (Blumenthal *et al.*, 2009).

In this chapter, we presented a first suggestion of how the robustness of a literature analysis with respect to an extended 'Levins' space' (i.e. the trade-off between precision, generality and realism, see Chapters 4 and 6, this volume, for explanation) could be tested. The approach we chose here is easy to implement; however, several aspects of this method deserve closer consideration. For example, the question is whether precision is best accounted for by recording the number of replicates of a study. With this approach, observations tend to receive a lower score for precision, as usually several samples per site in an observational study are viewed as pseudoreplication (and therefore were not counted as replicates), whereas several blocks in an experiment are viewed as true replicates, even if they are positioned in the same area. Also, our approach reflects multi-species studies not the way they are usually intended. In our analysis, those studies received low numbers for precision and high number for generality because here species are not counted as replicates. Despite these and other open questions, we believe that it is very helpful to test the robustness of data based on 'Levins' space', and thus to test whether the way studies are designed and performed affects their result. With this first suggestion, we hope to stimulate future discussions and research on this topic.

Conclusions

On the basis of our results, we suggest that it is high time to refine the enemy release hypothesis. Because it is not clear in which cases low levels of damage from or infestation with enemies are really linked to enhanced performance, future studies should focus on the performance of species with low pressure from enemies. A reformulation of the enemy release hypothesis could therefore read: 'invaders show enhanced performance if released from specialist enemies'.

For all future studies testing the enemy release hypothesis, it could be promising to include time since introduction as a factor because it has been shown that the longer an invader interacts with predators in the invaded range, the more resident predators are able to use it as a host or prey organism (Kroft and Blakeslee, 2016; Schilthuizen *et al.*, 2016). To avoid a bias in the selection of focal species, we suggest using a large set of study species (ten or more). Because results of studies in the lab and enclosure in our analysis differed from those done in the field, we further suggest that future studies are done in the field if feasible. Studies in laboratories and enclosures are certainly well suited if the aim is to reveal underlying mechanisms but if the aim is to test whether enemy release is enhancing invader success under realistic conditions, we believe field studies are more appropriate. For future synthesis, the most promising way forward seems to be to think of invasion successes and failures as complex phenomena, which in some cases might be partly driven by enemy release. The aim of future studies should be to deduce further in which cases a release from specialist enemies has an influence on invasive species performance and thus to determine the range of applicability of the enemy release hypothesis.

References

Blumenthal, D., Mitchell, C.E., Pyšek, P. and Jarošik, V. (2009) Synergy between pathogen release and resource availability in plant invasion. *Proceedings of the National Academy of Sciences of the United States of America* 106, 7899–7904.

Cappuccino, N. and Carpenter, D. (2005) Invasive exotic plants suffer less herbivory than non-invasive exotic plants. *Biology Letters* 1, 435–438.

Clark, N.J., Olsson-Pons, S., Ishtiaq, F. and Clegg, S.M. (2015) Specialist enemies, generalist weapons and the potential spread of exotic pathogens: malaria parasites in a highly invasive bird. *International Journal for Parasitology* 45, 891–899.

Colautti, R.I., Ricciardi, A., Grigorovich, I.A. and MacIsaac, H.J. (2004) Is invasion success explained by the enemy release hypothesis? *Ecology Letters* 7, 721–733.

Enders, M., Hütt, M.-T. and Jeschke, J.M. (2018) Drawing a map of invasion biology based on a network of hypotheses. *Ecosphere.* DOI: 10.1002/ecs2.2146.

Heger, T. and Jeschke, J.M. (2014) The enemy release hypothesis as a hierarchy of hypotheses. *Oikos* 123, 741–750.

Heger, T., Pahl, A.T., Botta-Dukat, Z., Gherardi, F., Hoppe, C., Hoste, I., Jax, K., Lindström, L., Boets, P., Haider, S. *et al.* (2013a) Conceptual frameworks and methods for advancing invasion ecology. *Ambio* 42, 527–540.

Heger, T., Saul, W.-C. and Trepl, L. (2013b) What biological invasions 'are' is a matter of perspective. *Journal for Nature Conservation* 21, 93–96.

Jeschke, J.M. (2014) General hypotheses in invasion ecology. *Diversity and Distributions* 20, 1229–1234.

Jeschke, J.M., Gómez Aparicio, L., Haider, S., Heger, T., Lortie, C.J., Pyšek, P. and Strayer, D.L. (2012) Support for major hypotheses in invasion biology is uneven and declining. *NeoBiota* 14, 1–20.

Keane, R.M. and Crawley, M.J. (2002) Exotic plant invasions and the enemy release hypothesis. *Trends in Ecology & Evolution* 17, 164–170.

Kowarik, I. and Pyšek, P. (2012) The first steps towards unifying concepts in invasion ecology were made one hundred years ago: revisiting the work of the Swiss botanist Albert Thellung. *Diversity and Distributions* 18, 1243–1252.

Kroft, K.L. and Blakeslee, A.M.H. (2016) Comparison of parasite diversity in native panopeid mud crabs and the invasive Asian shore crab in estuaries of northeast North America. *Aquatic Invasions* 11, 287–301.

Lafferty, K.D., Torchin, M.E. and Kuris, A.M. (2010) The geography of host and parasite invasions. In: Morand, S. and Krasnov, B.R. (eds) *The Biogeography of Host-parasite Interactions.* Oxford University Press, Oxford, UK, pp. 191–203.

Liu, H. and Stiling, P. (2006) Testing the enemy release hypothesis: a review and meta-analysis. *Biological Invasions* 8, 1535–1545.

Liu, H., Stiling, P., Pemberton, R.W. and Pena, J. (2006) Insect herbivore faunal diversity among invasive, non-invasive and native *Eugenia* species: Implications for the enemy release hypothesis. *Florida Entomologist* 89, 475–484.

Lowry, E., Rollinson, E.J., Laybourn, A.J., Scott, T.E., Liello-Lammens, M.E., Gray, S.M., Mickley, J. and Gurevitch, J. (2013) Biological invasions: a field synopsis, systematic review, and database of the literature. *Ecology and Evolution* 3, 182–196.

Merella, P., Pais, A., Follesa, M.C., Farjallah, S., Mele, S., Piras, M.C. and Garippa, G. (2016) Parasites and Lessepsian migration of *Fistularia commersonii* (Osteichthyes, Fistulariidae): shadows and light on the enemy release hypothesis. *Marine Biology* 163, 97.

Mitchell, C.E. and Power, A.G. (2003) Release of invasive plants from fungal and viral pathogens. *Nature* 421, 625–627.

Pohlert, T. (2014) The Pairwise Multiple Comparison of Mean Ranks Package (PMCMR). R package, Available at: http://CRAN.R-project.org/package=PMCMR (accessed 6 October 2017).

R Core Team (2014) R: A language and environment for statistical computing. R Foundation for Statistical Computing, Vienna, Austria. Available at: http://www.R-project.org/ (accessed 6 October 2017).

Saul, W.-C., Jeschke, J.M. and Heger, T. (2013) The role of eco-evolutionary experience in invasion success. *NeoBiota* 17, 57–74.

Schilthuizen, M., Pimenta, L.P.S., Lammers, Y., Steenbergen, P.J., Flohil, M., Beveridge, N.G.P., van Duijn, P.T., Meulblok, M.M., Sosef, N., van de Ven, R. *et al.* (2016) Incorporation of an invasive plant into a native insect herbivore food web. *PeerJ* 4, e1954.

Torchin, M.E., Lafferty, K.D., Dobson, A.P., McKenzie, V.J. and Kuris, A.M. (2003) Introduced species and their missing parasites. *Nature* 421, 628–630.

12 Evolution of Increased Competitive Ability and Shifting Defence Hypotheses

Caroline Müller*

Department of Chemical Ecology, Bielefeld University, Bielefeld, Germany

Abstract

This chapter focuses on two of the various hypotheses that take into account the role of biotic interactions in invasion biology, namely the evolution of increased competitive ability (EICA) hypothesis and the shifting defence hypothesis (SDH). Both hypotheses mainly consider changes in concentrations of chemical defences in plant individuals from native vs exotic populations and are studied independently of the novelty of the chemical metabolite in the invasive range. The EICA hypothesis predicts that chemical defences should be lower in plants of invasive populations because enemy pressure is reduced in the exotic range. The SDH predicts a shift in chemical defences by discriminating between qualitative and quantitative defences. Qualitative defences (toxins) are cheaper to produce and expected to increase in invasive populations because they are needed as defence against generalists, whereas expensive quantitative defences (digestibility reducers) may be reduced in plants of the invasive range owing to the overall lower enemy pressure. Methodological issues are pointed out that should be considered when testing the differences in plant traits between native and invasive populations. A literature review on qualitative and quantitative defence traits, which were compared between plants of native and invasive origin grown under standardized common-garden conditions, revealed 37 studies, comprising 22 plant species. The results of the review infrequently support predictions for defence distributions of the EICA hypothesis, whereas predictions of the SDH are supported by somewhat more traits. The definition of qualitative vs quantitative defences has shortcomings, however. In particular, actual costs of these defences are difficult to estimate but should be investigated in future studies. Furthermore, instead of focusing on individual defences, multiple defences should be considered in plant species that are invasive to gain a more complete understanding of resource-allocation patterns.

Introduction

In this chapter, first the basic concept of the evolution of increased competitive ability (EICA) hypothesis is introduced, which deals with plant traits regarding growth and defence that change throughout space and time. When researchers study the predictions of the EICA hypothesis or in general compare traits of native and exotic plant populations, several methodological aspects have to be taken into account, which are outlined in the second part of this chapter. Within this part, recommendations are

* Corresponding author. E-mail: caroline.mueller@uni-bielefeld.de

given for the set-up of respective experiments and the choice of relevant growth and defence traits. One main aspect of the EICA hypothesis is the investment into chemical defence. Defence traits can, however, be distinguished in different categories, for example, cheap qualitative vs expensive quantitative defences. Because of their distinct costs in terms of resource investment, these types of defences should respond in different ways under the release of enemies in the exotic range, which is considered in a refinement of the EICA hypothesis, the shifting defence hypothesis (SDH). The SDH is explained in more detail in the third part of this chapter. In the fourth part, a recent literature review is presented, in which I compiled all studies published on qualitative vs quantitative defence traits examined in plants of native and invasive populations grown under common conditions. The mean differences in investment in qualitative or quantitative defences in native vs invasive populations give indications of how far predictions of the EICA hypothesis vs the SDH can be supported. The distinction into these types of defences has some shortcomings, however, which are highlighted in the fifth part of this chapter before some general conclusions are drawn.

Evolution of Increased Competitive Ability Hypothesis

The EICA hypothesis was framed more than 20 years ago (Blossey and Nötzold, 1995) and is based on three observations or hypotheses: (i) in exotic environments, plants are often more vigorous and produce more biomass and more seeds compared to the respective phenotypes in the native environments (Noble, 1989); (ii) the enemy release hypothesis predicts that the absence (or at least reduction) of enemies in the invaded range is a cause of the invasion success (Chapter 11, this volume; Keane and Crawley, 2002; Mitchell and Power, 2003); (iii) the optimal defence hypothesis postulates that plants invest their resources in growth vs defence depending on the value of

the tissue to maximize individual fitness (McKey, 1979; Strauss et al., 2004). If resources are limited, plants will face trade-offs in resource allocation to growth, storage, reproduction, chemical or structural defences (Coley et al., 1985; Bazzaz et al., 1987; Herms and Mattson, 1992). On the basis of these concepts, it can be expected that, in alien plant populations, selection should favour and maintain genotypes with high competitive abilities (i.e. improved vegetative growth, high reproductive output) but low allocation in enemy defences. With regard to the latter, Blossey and Nötzold (1995) made the assumption that specialist herbivores should show an improved performance on plant individuals from invasive populations. Thus, originally the EICA hypothesis was restricted to herbivore defence but may likewise be extended to enemy defence in general (i.e. herbivores and pathogens). Furthermore, while Blossey and Nötzold (1995) focused on the consequences of plant chemical changes on the herbivores, the underlying mechanisms involved in herbivore defence, i.e. the composition and concentration of allelochemicals, were not discussed in the original paper. Only later publications testing the EICA hypothesis also measured concentrations of plant metabolites, predicting that concentrations of plant defence metabolites should be lower in plants of invasive origin (Table 12.1; discussed in detail below).

Changes in time and space

The evolution of adapted phenotypes in the context of plant invasions cannot be seen just as black and white, i.e. vigorously growing, poorly defended plants in the invasive range vs small but well-defended plants in the native range. Instead, adaptations to the environment are probably undergoing a transitory process. The first plant populations that are introduced to another continent may not be recognized and exploited by native herbivores, leading to a selection of plants that allocate more to growth and less to defence (Chew and Courtney, 1991).

Table 12.1. Common garden studies that investigated qualitative, quantitative and indirect defence traits in plants of native and invasive populations.

Species	Family	Trait	Ratio trait value invasive/native	Sign. (Y: p<0.05, N: p>0.05)	Tissue	No. invasive populations	Invasive area	No. native populations	Native area	Location of common garden	Further notes	References
(a) Qualitative defences (toxins)												
Melaleuca quinquenervia	Myrtac.	1,8-Cineole	0.51	Y	Leaf	10	USA	8	Australia, New Caledonia	Invasive	Three terpenoids picked out from 20	Franks et al., 2012
Ulex europaeus	Fabac.	Quinolizidine alkaloids	0.54	N	Shoots	6	New Zealand, Reunion	6	Europe	Native		Hornoy et al., 2012
Alliaria petiolata	Brassicac.	Total glucosinolates	0.61	N	Leaf (not induced)	3	North America	7	Europe	Invasive		Cipollini et al., 2005
Hypericum perforatum	Hypericac.	Hypericin	0.67	Y	Leaf	26	North America	16	Europe	Native		Maron et al., 2004a
Rorippa austriaca	Brassicac.	Glucosinolates	0.71	N	Leaf	7	The Netherlands, Germany	5	Czech Republic	Invasive	Range expander, within continent	Huberty et al., 2014
Solidago gigantea	Asterac.	Total diterpenes	0.72	Y	Leaf	22	Europe	10	USA	Invasive	Eight diterpenoids analysed	Johnson et al., 2007
Verbascum thapsus	Scrophulariac.	Iridoid glycoside: aucubin	0.73	N	Leaf (young)	6	America	6	Europe	Invasive	Sign. continent×leaf age interaction	Alba et al., 2012
Hypericum perforatum	Hypericac.	Hypericin	0.81	N	Leaf	32	North America	18	Europe	Invasive		Maron et al., 2004a
Hypericum perforatum	Hypericac.	Hypericide	0.81	Y	Leaf	32	North America	18	Europe	Invasive		Maron et al., 2004a
Solidago gigantea	Asterac.	Diterpenes	0.87	N	Leaf (undamaged)	20	Europe	10	USA	Native		Hull-Sanders et al., 2007
Solidago gigantea	Asterac.	Sesquiterpenes	0.92	N	Leaf (undamaged)	20	Europe	10	USA	Native		Hull-Sanders et al., 2007
Verbascum thapsus	Scrophulariac.	Iridoid glycoside: catalpol	0.95	N	Leaf (young)	6	America	6	Europe	Invasive	Sign. continent×leaf age interaction	Alba et al., 2012
Solidago gigantea	Asterac.	Total sesquiterpenes	1.00	N	Leaf	22	Europe	10	USA	Invasive	Seven sesquiterpenoids identified	Johnson et al., 2007
Cynoglossum officinale	Boraginac.	Pyrrolizidine alkaloids	1.01	N	Leaf (constitutive)	3	North America	4	Europe	Invasive		Eigenbrode et al., 2008
Lepidium draba	Brassicac.	Total glucosinolates (GS)	1.05	N	Mature leaves	10	North America	11	Europe	Native	Sign. differences for individual GS	Müller and Martens, 2005

continued

Table 12.1. *continued*

Species	Family	Trait	Ratio trait value invasive/native	Sign. (Y: $p<0.05$, N: $p>0.05$)	Tissue	No. invasive populations	Invasive area	No. native populations	Native area	Location of common garden	Further notes	References
Lepidium draba	Brassicac.	Total glucosinolates	1.07	N	Cotyledons and first true leaves	10	North America	11	Europe	Native		Müller and Martens, 2005
Melaleuca quinquenervia	Myrtac.	Viridiflorol	1.10	N	Leaf	10	USA	8	Australia, New Caledonia	Invasive	Three terpenoids picked out from 20	Franks *et al.*, 2012
Asclepias syriaca	Asclepiadac.	Cardenolides	1.29	Y	Leaf (control)	10	Europe	10	North America	Native		Agrawal *et al.*, 2015
Melaleuca quinquenervia	Myrtac.	E-nerolidol	1.34	Y	Leaf	10	USA	8	Australia, New Caledonia	Invasive	Three terpenoids picked out from 20	Franks *et al.*, 2012
Brassica nigra	Brassicac.	Glucosinolate: sinigrin	1.38	Y	Leaf	8	North America	16	Europe, Africa, Asia	Native		Oduor *et al.*, 2011
Tanacetum vulgare	Asterac.	Mono- and sesquiterpenes	1.41	Y	Leaf	9	North America	13	Europe	Native		Wolf *et al.*, 2011
Triadica sebifera[a]	Euphorbiac.	Flavonoids (total of five flavonoids)	1.44	Y	Young leaves of seedlings (control)	8	North America	8	China	Native		Wang *et al.*, 2012
Lepidium draba	Brassicac.	Soluble myrosinase activities	1.48	Y	Mature leaves	10	North America	11	Europe	Native		Müller and Martens, 2005
Jacobaea vulgaris[b]	Asterac.	Pyrrolizidine alkaloids	1.56	Y	Leaf	26	New Zealand, North America	6	Europe	Native		Rapo *et al.*, 2010
Asclepias syriaca	Asclepiadac.	Salicylic acid	1.68	N	Leaf (control)	10	Europe	10	North America	Native	Induced levels of SA sign. different	Agrawal *et al.*, 2015
Centaurea maculosa	Asterac.	Catechin	1.75	N	Seedlings	11	North America	4	Europe	Invasive		Ridenour *et al.*, 2008
Senecio jacobaea[b]	Asterac.	Pyrrolizidine alkaloids	1.83	Y	Leaf	4	New Zealand, USA	4	Europe	Native		Stastny *et al.*, 2005
Senecio jacobaea[b]	Asterac.	Pyrrolizidine alkaloids	1.89	Y	Leaf	16	North America/ Australia/New Zealand	15	Europe	Native		Joshi and Vrieling, 2005
Centaurea maculosa	Asterac.	Phytol	2.16	Y	Leaf	23	North America	22	Europe	Invasive		Ridenour *et al.*, 2008
Alliaria petiolata	Brassicac.	Isovitexin 6'''-O-β-D-glucopyranoside	2.19	Y	Leaf (not induced)	4	North America	7	Europe	Invasive		Cipollini *et al.*, 2005

Species	Family	Defence trait	Value	Y/N	Tissue	n	Range	n	Range	Status	Notes	Reference
Chromolaena odorata	Asterac.	Odoratin (chalcone)	2.40	Y	Leaf (newly mature)	10	Asia	10	America	Invasive		Zheng *et al.*, 2015
Centaurea maculosa	Asterac.	Germacrene D	2.92	N	Leaf	23	North America	22	Europe	Invasive	Response to JA varied by continent	Ridenour *et al.*, 2008
Alliaria petiolata	Brassicac.	Alliarinoside	3.37	N	Leaf (not induced)	4	North America	7	Europe	Invasive		Cipollini *et al.*, 2005
Senecio pterophorus	Asterac.	Pyrrolizidine alkaloids	4.45	Y	Leaf (control)	3	Europe	3	South Africa	Invasive		Caño *et al.*, 2009
Senecio inaequidens	Asterac.	Pyrrolizidine alkaloids	817.00	Y	Leaf (control)	3	Europe	3	South Africa	Invasive		Caño *et al.*, 2009
(b) Quantitative defences (digestibility-reducers)												
Sapium sebiferum[a]	Euphorbiac.	Tannins	0.24	Y	Leaf		North America		China			Siemann and Rogers, 2003
Triadica sebifera[a]	Euphorbiac.	Tannins	0.50	Y	Young leaves of seedlings (control)	8	North America	8	China	Native	Total of four tannins	Wang *et al.*, 2012
Ageratina adenophora	Asterac.	Cell wall protein content	0.57	Y	Leaf	10	China/India	5	Mexico	Invasive		Feng *et al.*, 2009
Persicaria perfoliata	Polygonac.	Tannins	0.65	Y	Leaf	4	North America	3	East Asia	Native		Guo *et al.*, 2011
Lythrum salicaria	Lythrac.	Total phenolics	0.69	Y	Leaf	6	North America/Australia	6	Europe	Native		Willis *et al.*, 1999
Lepidium draba	Brassicac.	Polyphenolics	0.73	Y	Mature leaves	10	North America	11	Europe	Native		Müller and Martens, 2005
Triadica sebifera[a]	Euphorbiac.	Tannins	0.78	Y	Leaf of 1-year-old seedlings	6	South-east USA	6	China	Native		Huang *et al.*, 2010
Triadica sebifera[a]	Euphorbiac.	Tannins	0.79	Y	Leaf of 2-year-old seedlings	6	South-east USA	6	China	Native		Huang *et al.*, 2010
Triadica sebifera[a]	Euphorbiac.	Tannins	0.83	Y	Leaf of 3-year-old seedlings	6	South-east USA	6	China	Native		Huang *et al.*, 2010
Senecio jacobaea[b]	Asterac.	Dry matter content	0.85	Y	Leaf	14	North America/Australia/New Zealand	8	Europe	Native		Doorduin *et al.*, 2011
Asclepias syriaca	Asclepiadac.	Trichomes	0.88	N	Leaf	10	Europe	10	North America	Native		Agrawal *et al.*, 2015
Jacobaea vulgaris[b]	Asterac.	Cell wall proteins	0.89	N	Leaf	20	New Zealand, Australia, USA	19	Europe	Native		Lin *et al.*, 2015
Chromolaena odorata	Asterac.	Toughness	0.93	Y	Leaf	8	Asia	8	America	Invasive	High nutrients	Liao *et al.*, 2013
Silene latifolia	Caryophyllac.	Trichomes	0.94	N	Calyx	20	North America	20	Europe	Invasive		Blair and Wolfe, 2004

continued

Table 12.1. *continued*

Species	Family	Trait	Ratio trait value invasive/native	Sign. (Y: $p < 0.05$, N: $p > 0.05$)	Tissue	No. invasive populations	Invasive area	No. native populations	Native area	Location of common garden	Further notes	References
Chromolaena odorata	Asterac.	Toughness	0.94	N	Leaf	8	Asia	8	America	Invasive	Low nutrients	Liao *et al.*, 2013
Jacobaea vulgaris[b]	Asterac.	Leaf thickness	0.96	N	Leaf	20	New Zealand, Australia, USA	19	Europe	Native		Lin *et al.*, 2015
Spartina alterniflora[c]	Poac.	Cell wall	1.09	N	Leaf	3	China	5	North America	Native and invasive	Low nitrogen, nitrogen had sign. effect	Qing *et al.*, 2012
Spartina alterniflora[c]	Poac.	Cell wall	1.10	N	Leaf	3	China	5	North America	Native and invasive	High nitrogen, nitrogen had sign. effect	Qing *et al.*, 2012
Chromolaena odorata	Asterac.	Total phenolics	1.02	N	Leaf	8	Asia	8	America	Invasive	High nutrients	Liao *et al.*, 2013
Alliaria petiolata	Brassicac.	Trypsin inhibitors	1.04	N	Leaf (not induced)	4	North America	7	Europe	Invasive		Cipollini *et al.*, 2005
Silene latifolia	Caryophyllac.	Trichomes	1.05	N	Leaf	20	North America	20	Europe	Invasive		Blair and Wolfe, 2004
Jacobaea vulgaris[c]	Asterac.	Toughness	1.08	N	Leaf	20	New Zealand, Australia, USA	19	Europe	Native		Lin *et al.*, 2015
Chromolaena odorata	Asterac.	Total phenolics	1.13	Y	Stems	8	Asia	8	America	Invasive		Zheng *et al.*, 2013
Alliaria petiolata	Brassicac.	Peroxidase activity	1.15	N	Leaf (not induced)	4	North America	7	Europe	Invasive		Cipollini *et al.*, 2005
Solidago gigantea	Asterac.	Short-chain hydocarbons	1.15	N	Leaf (undamaged)	20	Europe	10	USA	Native		Hull-Sanders *et al.*, 2007
Chromolaena odorata	Asterac.	Total phenolics	1.22	Y	Leaves	8	Asia	8	America	Invasive		Zheng *et al.*, 2013
Chromolaena odorata	Asterac.	Total phenolics	1.22	Y	Leaf	8	Asia	8	America	Invasive	Low nutrients	Liao *et al.*, 2013
Asclepias syriaca	Asclepiadac.	Latex	1.35	N	Leaf (control)	10	Europe	10	North America	Native		Agrawal *et al.*, 2015

Species	Family	Defence trait	Ratio	Sign.	Tissue		Native range		Invasive range	Status	Reference
Chromolaena odorata	Asterac.	Trichomes	1.36	Y	Five leaves per plant (lower surface)	15	Asia	13	America	Invasive	Liao et al., 2014
Centaurea maculosa	Asterac.	Trichome	1.44	Y	Leaf	23	North America	22	Europe	Invasive	Ridenour et al., 2008
Persicaria perfoliata	Polygonac.	Prickle density	1.57	Y	Leaf	4	NA	3	East Asia	Native	Guo et al., 2011
Chromolaena odorata	Asterac.	Trichomes	2.20	Y	Five leaves per plant (upper surface)	15	Asia	13	America	Invasive	Liao et al., 2014
Centaurea maculosa	Asterac.	Toughness	2.66	Y	Leaf	23	North America	22	Europe	Invasive	Ridenour et al., 2008
Brassica nigra	Brassicac.	Trichomes	2.92	Y	Leaf	8	NA	16	Europe, Africa, Asia	Native	Oduor et al., 2011
(c) Indirect defences											
Triadica sebifera[a]	Euphorbiac.	Extrafloral nectaries (EFN)	0.42	Y	Leaf control	8	North America	8	China	Native	Wang et al., 2013
Triadica sebifera[a]	Euphorbiac.	Extrafloral nectaries	0.75	Y	Leaves (induced and control)	18	North America	20	China, Japan	Native	Carrillo et al., 2012
Triadica sebifera[a]	Euphorbiac.	Extrafloral nectaries	0.76	Y	Leaves (induced and control)	18	North America	20	China, Japan	Native	Carrillo et al., 2012
Triadica sebifera[a]	Euphorbiac.	Extrafloral nectaries	0.87	Y	Leaves (induced and control)	18	North America	20	China, Japan	Native	Carrillo et al., 2012
Triadica sebifera[a]	Euphorbiac.	Extrafloral nectaries	1.49	N	Leaf	11	North America	8	Asia	Invasive	Carrillo et al., 2014

Using the Web of Science, the literature was searched for publications including the terms 'common garden and invasive and defen*' or 'invasive and native plant* and defen*' published until August 2016. Only studies were included in which at least three populations per range were analysed. The 'ratio of the trait value invasive/native' is based on means of values shown in figures or mentioned in the text of the respective references. The column 'Sign.' (significance) states whether the authors found a significant difference between native and invasive populations for the respective trait (Y – yes, N – no).
[a] Sapium sebiferum and Triadica sebifera are synonymous.
[b] Senecio jacobaea and Jacobaea vulgaris are synonymous.
[c] Counted as one trait because only nitrogen conditions differed.
JA, jasmonic acid; SA, salicylic acid.

Native herbivores might, however, start to explore the invasive plants as a highly abundant and edible resource. As a consequence, plants might be selected to allocate more into defence again (Chew and Courtney, 1991; Adler, 1999). Empirical evidence for such a transition of defences, i.e. a loss and reacquisition, has been gathered, for example, for *Sapium sebiferum* (Euphorbiaceae) that colonized another continent at different time points (Siemann and Rogers, 2001). In native populations of this species, tannin concentrations were highest, whereas in genotypes of recently invaded areas (early 20th century) they were not detectable but plants showed the highest growth. In contrast, in *S. sebiferum* plants of areas already colonized in the late 18th century the tannin concentrations were intermediate (Siemann and Rogers, 2001).

Next to cross-continental expansion, recent intra-continental range expansion (within about the last 100 years) caused by direct or indirect anthropogenic processes can also lead to plant phenotypes that exhibit invasive properties (Engelkes *et al.*, 2008; Fortuna *et al.*, 2014). These intra-continental range expanders may encounter at least some of the herbivore or pathogen community that they also face in their native distribution area, although the associated above- and below-ground species might not migrate at the same rate as their host plants (Morriën *et al.*, 2010). Moreover, intra-continental range-expanders might still experience ongoing gene-flow between newly established and native populations (Morriën *et al.*, 2010). Nevertheless, these range-expanders may evolve modifications of phenotypic traits including plant chemistry (Fortuna *et al.*, 2014).

Furthermore, traits differ between invasive plants growing behind the invasion front (core subpopulations) and along invasion fronts (edge subpopulations) (Rice *et al.*, 2013), which can be caused by a combination of adaptive genetic variation and plastic responses on a very local scale in subpopulations with a reduced gene flow (Dietz and Edwards, 2006). Variation can also occur in the enemy pressure across invasive populations. Such population-specific selection pressures lead to geographic mosaics of decreased and increased defence investment in various species (geographic mosaic of coevolution; Thompson, 1999; Orians and Ward, 2010). For example, *Brassica oleracea* plants exposed to higher herbivore pressure show increased concentrations of glucosinolates (Mithen *et al.*, 1995). Such among-population differences in the local selection pressures on defence investment should thus be considered when comparing defence concentrations in plants of different origin.

Methodological Considerations

To test the predictions of the EICA hypothesis, some methodological issues have to be considered when investigating differences in phenotypes of plants belonging to native vs invasive populations. First, the location of the experiment and the choice of the experimental material can be crucial. Second, the traits that are investigated with regard to plant vigour, competitive ability and defence need to be chosen with care.

Choice of appropriate study designs

To compare growth and defence-related traits, plants originating from different source populations of the native and invasive range should be grown under standardized, so-called common-garden conditions. In this way, the genetically based phenotypic differentiation among populations can be quantified and the influence of environmental variation is reduced (Blossey and Nötzold, 1995; Colautti *et al.*, 2009). Furthermore, investigating as many populations as possible allows more general conclusions and increases the statistical power. The maximum number of populations, in which defence concentrations have been measured in a common garden, is 32 for populations of the invasive origin and 50 for the total number of native and invasive populations (Maron *et al.*, 2004a; Table 12.1). This might not always be feasible but a minimum of at least five populations per range may be recommended.

The location of the common garden itself can also be crucial. In most studies, populations are grown in one common garden that is either located in the native or the invasive range of the species' distribution, often simply due to logistic reasons. Only few studies used replicated gardens (e.g. Maron et al., 2004a; Qing et al., 2012). Such replicated gardens allow investigating genotype-by-environment (G×E) interactions and, indeed, in most studies using this experimental design evidence for G×E effects have been revealed (Colautti et al., 2009). The garden location probably favours local genotypes (Kleine et al., 2017), which should at least be kept in mind when interpreting data from single garden studies. An important source causing differences in defence chemistry between native and invasive populations grown in different environments could be the soil type as well as the microbiota associated with that soil, mediating different plant–soil feedbacks (Callaway et al., 2004; Schweitzer et al., 2014). Although the association between soil microbiota (e.g. arbuscular mycorrhizal fungi) and plants is often morphologically restricted to the roots, this can also drastically influence leaf chemical composition owing to exchange of nutrients and phytohormone signalling (Schweiger and Müller, 2015).

Furthermore, sources of among-population variation need to be taken into account when comparing phenotypic traits (Colautti et al., 2009). For example, plant height or reproductive output correlate with latitude, indicating local adaptation to latitudinal clines, i.e. differences in climate, growing season and day length (Colautti et al., 2009), but also in herbivore pressure (Orians and Ward, 2010). Therefore, it is highly recommended to include latitude as a factor in the statistical data evaluation (Maron et al., 2004b; Colautti et al., 2009; Wolf et al., 2011; Kleine et al., 2017). Ideally, native and invasive populations used for a common-garden study should be collected from matching latitudes (Colautti et al., 2009) but this might not always be feasible.

Moreover, the magnitude and direction of growth–defence trade-offs probably depend on environmental variables (Schuman and Baldwin, 2016). Surprisingly few studies have included the role of resource availability in common garden experiments (but see e.g. Qing et al., 2012; Liao et al., 2013). Under limited nitrogen conditions, invasive plants of Spartina alterniflora (Poaceae) showed an increased allocation of leaf nitrogen to ribulose-1,5-bisphosphate-carboxylase/-oxygenase (RuBisCO) compared to plants from native populations (Qing et al., 2012). Differences in water availability had an effect on growth and regrowth abilities of Tanacetum vulgare (Asteraceae) but there was no significant drought treatment×plant origin effect that would indicate differences in phenotypic plasticity (Kleine et al., 2017). More studies are needed that implement the relevance of resource availability in their study design.

A high variation in traits between native and invasive populations is not necessarily due to different selection pressures and evolution of adapted phenotypes. Next to founder events, genetic drift, hybridization or bottlenecks after introduction (Keller and Taylor, 2008), environmental maternal or epigenetic effects are potential alternative explanations. Common garden experiments using field-collected seeds may lead to inaccurate results due to environmental maternal effects, particularly for traits important in early stages of development (Roach and Wulff, 1987; Bossdorf et al., 2005). For example, larger seeds may have higher viability and germinate faster. Thus, seed size or other growth traits of the mother plant should be included as covariate (Bossdorf et al., 2005) to account for environmental effects that led to a certain plant phenotype at a particular point in space and time. In species with extremely small seeds, such carry-over effects may be negligible (Blossey and Nötzold, 1995). Overall, maternal effects might be insignificant in some plant species (e.g. Monty et al., 2009) but could play an important role in others and might depend on the environmental conditions (Campbell *et al.*, 2015).

To exclude the influence of maternal effects, generating an F1 or subsequent generations under standardized conditions that are then used for common-garden studies is recommended (Bossdorf *et al.*, 2005). However, whereas propagating new generations is easily feasible in selfing and in annual plants, for outcrossing plants it needs to be ensured that cross-fertilization only occurs within each population. Moreover, for biannual plants it takes some time until new seeds can be harvested.

Exposure to distinct environmental conditions can also lead to rapid genome-wide epigenetic reprogramming, such as methylation alteration, leading to acclimation (Gao *et al.*, 2010). It is, however, relatively unexplored to what extent epigenetic modification can facilitate invasions by affecting resource allocation into growth vs defence and whether differences in epigenetic patterns are the cause or the consequence of habitat differentiation (Richards *et al.*, 2012; Gillies *et al.*, 2016). One way to test for the role of epigenetic variation is the use of demethylating agents, such as 5-azacytidine (Bossdorf *et al.*, 2008). However, the agent needs to be applied repeatedly and the effectiveness should be tested using molecular approaches.

Choice of traits to estimate plant vigour and competitive ability

To study the assumptions of the EICA hypothesis, various traits related to plant vigour and fitness, such as biomass, plant height, number of ramets (vegetative) or seed output (generative reproduction), are usually compared between plants of native and invasive origins that are grown individually in pots under standardized conditions. In line with the EICA hypothesis, in some species plants of the invasive range grow bigger than their native counterparts (Blossey and Nötzold, 1995; Buschmann

et al., 2005; Abhilasha and Joshi, 2009), which might be explained by a higher photosynthetic capacity (Lei *et al.*, 2011; Qing *et al.*, 2012). In contrast, in other species plants of native and invasive populations grow equally (Buschmann *et al.*, 2005; Müller and Martens, 2005) and even a reduced plant vigour has been revealed in invasive populations in a few species (Hinz and Schwarzlaender, 2004). These different outcomes can be explained by shifts in resource allocation, for example, from vegetative growth to sexual reproduction, which could lead to contrasting results for different plant traits (Hinz and Schwarzlaender, 2004).

To test for 'competitive ability', plants should be grown in competition but, surprisingly, this aspect of the EICA hypothesis has rarely been addressed. Competition experiments are performed using intraspecific and/or interspecific competition, which can lead to distinct results. For instance, invasive *Lythrum salicaria* (Lythraceae) only exhibited greater competitive effects and responses when grown together with interspecific neighbours but not when grown in intraspecific competition (Joshi *et al.*, 2014). Populations of *Alliaria petiolata* (Brassicaceae) showed no differences in growth and reproduction when grown alone but native populations outperformed the invasive ones when grown in competition with conspecifics (Bossdorf *et al.*, 2004). This observation led to the 'evolution of reduced competitive ability hypothesis' (Bossdorf *et al.*, 2004). Such a reduced competitive ability might be a consequence of potential genetic bottlenecks during invasion with subsequent inbreeding depression or genetic drift (van Kleunen and Schmid, 2003). Alternatively, directional selection due to distinct communities with fewer or weaker competitors in the invasive range in combination with costs involved in competition might lead to invasive phenotypes with a reduced competitive ability (Bossdorf *et al.*, 2004). A recent study testing the EICA hypothesis did not only consider single but also more realistic multi-species competition scenarios (Oduor *et al.*, 2015). The authors provide evidence that plants of native *Brassica nigra* (Brassicaceae) populations compete better than plants of invasive origin in the presence of a single strong interspecific competitor but the invasive populations are better competitors in a multi-species community (Oduor *et al.*, 2015). Future investigations should incorporate such multi-species scenarios to reveal the influence of competition under more natural conditions.

Studies on the competitive ability are performed either in environments free of herbivores (for example, using insecticides) or with shoot herbivore pressure (Franks *et al.*, 2008; Oduor *et al.*, 2011). The importance of root herbivory has only recently been considered in such studies (Oduor *et al.*, 2015; Zheng *et al.*, 2015). Root herbivory can reduce the negative effects of competition on the target plant, which may be due to compensatory growth (Oduor *et al.*, 2015). Furthermore, soil biota may act as important selective agents, as demonstrated in some tree species, where invasive plants cultivate soil biota that are beneficial for their offspring (Pregitzer *et al.*, 2010; Felker-Quinn *et al.*, 2011), leading to a positive plant–soil feedback.

The competitive superiority of invasive plant populations might undergo spatiotemporal changes. A recent meta-analysis revealed that negative competitive effects decline across sites that have been invaded for longer periods of time (Iacarella *et al.*, 2015). The pace of decline differs between plants of different functional groups, with negative competitive effects of invasive grasses declining more rapidly over time compared to invasive forbs, herbs and shrubs. These findings have considerable implications for the control of invasive plant species; it may be more important to prevent the spread of recent invaders than to control long established invaders owing to the spatio-temporal changes in their competitive abilities (Iacarella *et al.*, 2015). In conclusion, both the functional group and the invasion history can have implications for the expectations regarding changes in traits related to growth.

Choice and distinction in traits related to herbivore defence

Next to the effects on plant vigour and competitive ability, the defence capacity against herbivores is an important aspect of the EICA hypothesis. To study this capacity, a possibility is to focus on herbivore growth and survival, assuming that herbivores should perform less well on better defended native plants than on poorly defended invasive conspecifics, as indeed demonstrated in various studies (Blossey and Nötzold, 1995; Daehler and Strong, 1997). Alternatively, the focus can be on plant resistance, i.e. plant traits that reduce herbivore attack, or the tolerance of the plant, i.e. traits that allow the plant to maintain fitness even when being damaged (Oduor *et al.*, 2011).

Investment in defences may vary within an individual between plant tissues depending on their value (i.e. optimal defence), with higher defence levels often found in young compared to old leaves, particularly in plants of invasive populations (Alba *et al.*, 2012). Defence levels can also pronouncedly differ between seedlings and mature plants of invasive vs native populations (Müller and Martens, 2005). This spatial and temporal variation needs to be kept in mind when studying defence capabilities.

The plant traits causing reduced herbivory may be mechanical or chemical defences, which could be either constitutively expressed or locally or systemically induced in response to an attack by an antagonist (or a simulated attack using phytohormone treatments). In *A. petiolata*, lower constitutive concentrations but a higher inducibility of glucosinolates were found in plants of invasive compared to plants of native populations (Cipollini *et al.*, 2005). Induced pyrrolizidine alkaloid concentrations showed a significantly higher variance in invasive compared to native populations of *Cynoglossum officinale* (Boraginaceae), although the mean inducible levels were comparable (Eigenbrode *et al.*, 2008). The different responses in constitutive vs inducible defences highlight that care should be taken when generalizing the predictions of the

EICA hypothesis. More studies are needed that consider these differences in defences between native and invasive populations (Orians and Ward, 2010).

Inducible defences either act directly against the herbivore or mediate indirect defence, i.e. the attraction of natural enemies of the herbivores. If plants face low herbivore attack (as expected when growing in the invasive range), they might also invest less in induced indirect defences. In accordance with this assumption, invasive populations of *Triadica sebifera* (syn. *Sapium sebiferum*, Euphorbiaceae) show a reduced investment in extrafloral nectar production compared to their native conspecifics, supporting the EICA hypothesis for an indirect defence trait (Carrillo *et al.*, 2012). These results highlight that indirect defences involved in tritrophic interactions might be under a similar selection as direct defence traits in introduced populations.

Methodologically, it has been traditional to study one target metabolite or a group of metabolites belonging to the same chemical class that are characteristic for the target plant to investigate the EICA hypothesis. It is recommended, however, to investigate a larger set of defence-related traits within each individual because some of these traits may correlate negatively, even though no overall trade-off between defence traits in plants was found in a meta-analysis (Koricheva *et al.*, 2004). Large-scale metabolomics approaches have only recently started to be applied in invasion ecology but might reveal intriguing insights in changes of chemical fingerprints (Fortuna *et al.*, 2014; Macel *et al.*, 2014; Pankoke *et al.*, submitted; Tewes *et al.*, 2018). Apart from metabolite concentrations, activities of plant enzymes should also be considered that directly interfere with herbivore digestion or hydrolyse plant substrates, turning them into more toxic products (Cipollini *et al.*, 2005; Müller and Martens, 2005). Such more comprehensive studies take into account that different defence traits have distinct costs and that herbivores are differentially responsive towards them.

Furthermore, although most studies related to the EICA hypothesis focus on the effects on herbivores, only few studies include the effectiveness of defences against pathogens (but see e.g. Maron *et al.*, 2004a; Pankoke *et al.*, submitted). Metabolomics studies combined with bioassay-guided fractionation revealed that different compounds are responsible for resistance against leaf herbivores vs pathogens in the successful invader *Buddleja davidii* (Scrophulariaceae) (Pankoke *et al.*, submitted). Moreover, in most studies leaf chemistry is analysed but research is also needed on resistance traits of roots (Lankau, 2011).

A very important aspect to consider in relation to herbivore defence is to discriminate between specialists that feed on only one (monophagous) or a few species within a plant family (oligophagous) vs generalists that can potentially attack a wide range of plants belonging to several families (polyphagous). When spreading in a novel environment, exotic plants are expected to be, at least initially, released of specialist herbivores, whereas the pressure by generalist herbivores may vary. Thus, instead of experiencing a release from enemies, invasive species might rather experience a shift in herbivore community. This has important implications for plant defences, which have been taken into account in the 'shifting defence hypothesis' (SDH), as outlined below.

Refinement of the EICA Hypothesis by the Shifting Defence Hypothesis

The SDH (Müller-Schärer and Steinger, 2004) is based on the EICA hypothesis but incorporates considerations of the specialist–generalist dilemma (van der Meijden, 1996; Müller-Schärer and Steinger, 2004). Several of the small molecules that can be characteristic of a particular plant species repel or deter generalists (often referred to as 'toxins' or 'qualitative defences') but attract and stimulate specialists, causing a dilemma for the plant, because generalists and specialists thus exert contrasting selection pressures on defence concentrations. In the native habitat, plants are therefore expected to contain intermediate concentrations of these metabolites, which are readily produced, probably at low cost and acting at low concentrations (but see the discussion below on shortcomings). In the invasive range, plants should actually be able to increase the concentrations of such compounds to be well defended against generalists without running the risk of attracting specialists, which are usually lacking in that area (Müller-Schärer *et al.*, 2004; Fig. 12.1). This response would, however, be in contrast to predictions of the EICA hypothesis. Apart from acting as defence against generalists, these small molecules (e.g. glucosinolates

and their hydrolysis products) can be involved in allelopathic effects on neighbouring plants. Thus, increased concentrations of such allelochemicals might also benefit the invader in terms of competition (Müller, 2009).

In contrast to the qualitative defences, the 'quantitative defences' (also termed 'digestibility reducers'), such as lignins or tannins, are large molecules acting in large quantities. They reduce leaf palatability for both specialists and generalists and are assumed to be expensive for the plant, constraining its relative growth rate (Müller-Schärer and Steinger, 2004). Growth is

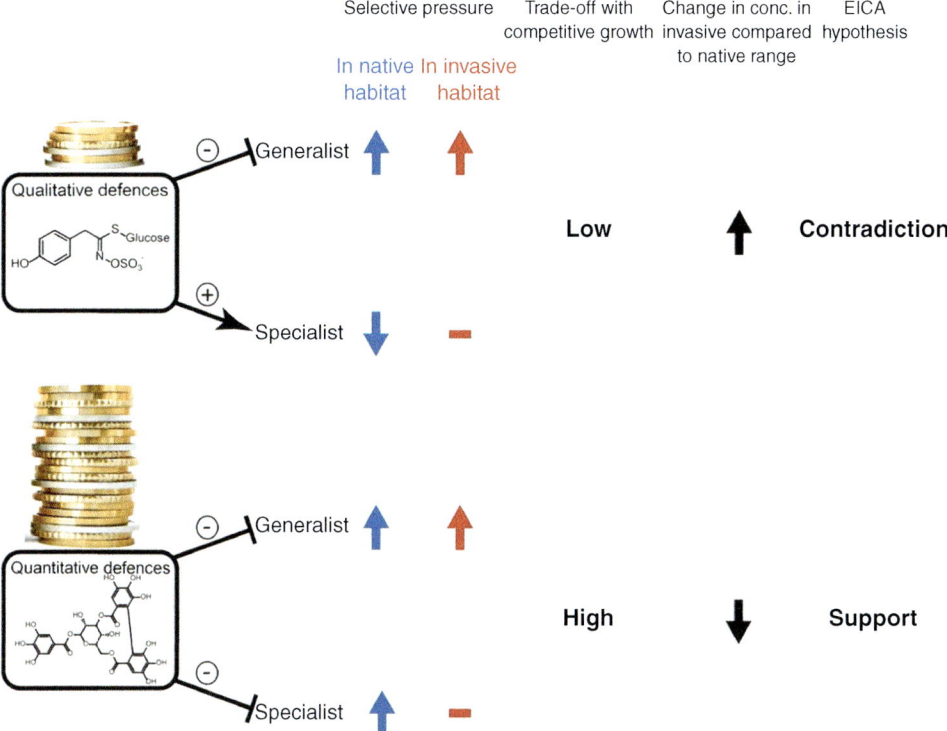

Fig. 12.1.   Selective pressures acting by specialists vs generalists on cheap toxins (qualitative defences, e.g. *p*-hydroxybenzyl glucosinolate) and expensive digestibility reducers (quantitative defences, e.g. gallotannin). In the native range of a species, opposite selection pressures act on concentrations of qualitative defences because these compounds fend off generalists but attract specialists. In the invasive range, where specialists are lacking, concentrations of cheap toxins should increase to keep generalists away (in contrast to predictions of the EICA hypothesis). Regarding quantitative defences, which act against both specialists and generalists, selection should lead to low concentrations in invasive conspecifics because these defences are expensive and the trade-off with competitive growth is high (in accordance with the EICA hypothesis). The differentiation in qualitative vs quantitative defences is considered in the shifting defence hypothesis.

correlated with competitive ability, however. Thus, in the native range of a plant, contrasting selection pressures by specialists and competing plants might again lead to an intermediate level of these defences. In the invasive range where overall a lower herbivore pressure is expected, selection pressure to produce expensive quantitative defences will be low and plants will instead invest in competitive growth (Müller-Schärer et al., 2004; Fig. 12.1), which is in line with predictions of the EICA hypothesis. Shifts in investment in qualitative rather than quantitative defences in invasive plants will result in a net gain of resources that can be invested in competitive growth.

Evaluation of Data on Defence Concentrations Supporting the EICA Hypothesis or the SDH

In a meta-study investigating the effects of leaf chemical and physical traits on herbivore performance in relation to the EICA hypothesis and the SDH (Felker-Quinn et al., 2013), plants of invasive populations tended, but not significantly so, to show higher defences against generalists and lower defences against specialists compared to plants of native origins, which would fit with the expected herbivore pressures in these ranges. In a review by Doorduin and Vrieling (2011), 15 publications were summarized in which concentrations of defences were analysed in plants reared in common gardens. For toxins (qualitative defences), results from nine species were in accordance with the predictions of the SDH (higher concentrations in invasive plants). Trichomes, toughness measures and dry matter content were considered as digestibility-reducing traits (quantitative defences) because these mechanical defences were grouped with digestibility reducers in one plant defence syndrome (Travers-Martin and Müller, 2008). For digestibility reducers, data for eight species revealed no significant decrease in plants from the invaded areas, in contrast to the predictions by SDH (Doorduin and Vrieling, 2011).

A recent literature survey, in which I screened the ISI Web of Science for studies including the terms 'common garden and invasive and defen*' or 'invasive and native plant* and defen*' published until August 2016, revealed several more studies (Table 12.1), also including all papers mentioned in earlier reviews on the EICA hypothesis or SDH (Hinz and Schwarzlaender, 2004; Doorduin and Vrieling, 2011; Felker-Quinn et al., 2013). In total, I found results for 22 plant species, belonging to 13 plant families, in 37 publications (Fig. 12.2). For both qualitative (chemical) and quantitative defences

(a) Qualitative defences

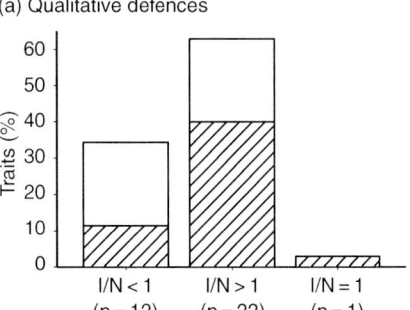

(b) Quantitative defences

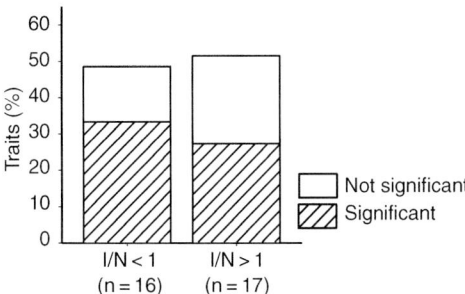

Fig. 12.2. Percentage of (a) qualitative defences (toxins) and (b) quantitative defences (digestibility reducers, including mechanical defences), which were found to be lower (I/N < 1), higher (I/N > 1) or equal (I/N = 1) in plants from invasive (I) compared to native (N) populations. Data are based on 37 publications, comprising 22 plant species, based on a literature search in ISI Web of Science for studies including the terms 'common garden and invasive and defen*' or 'invasive and native plant* and defen*' published until August 2016 (Table 12.1). Numbers (n) of traits are given in parentheses; significant differences for traits (p = 0.05) are indicated by hatched bars.

(mostly chemical but also other digestibility-reducing traits), a similar number of traits has been studied. For some plant species, several traits were investigated (up to five traits, e.g. Cipollini, 2005; Ridenour et al., 2008). All traits were considered individually in Table 12.1 and Fig. 12.2. For qualitative defences, more traits showed (significantly) higher (22 traits, from 14 plant species) than lower (12 traits, from nine species) concentrations in plants of invasive than native populations (Fig. 12.2, Table 12.1), which is in line with predictions of the SDH (Fig. 12.1). For quantitative defences, about the same number of traits was found to be (significantly) higher (17 traits, from ten plant species) or lower (16 traits, from nine plant species) in defence levels in plants of invasive vs native origin. This is in contrast to the SDH as well as the EICA hypothesis, which both predict lower defence levels in plants of the invasive range. In total, on the basis of this recent literature survey 41% of the analysed traits (28 out of 68, including qualitative and quantitative defences, Table 12.1 and Fig. 12.2) are in support of the EICA hypothesis, whereas somewhat more, namely 56% of the traits (22 out of 35 for qualitative defences and 16 out of 33 for quantitative defences), are in support of the SDH (including significant and not-significant differences in concentrations).

Interestingly, even within a plant species, quantitative defence traits can respond in contrasting directions, i.e. tannin concentrations were higher but prickle density was lower in plants of invasive populations compared to the native conspecifics in Persicaria perfoliata (Polygonaceae) (Guo et al., 2011; Table 12.1). This could potentially be an indication for trade-offs in defence investment in these expensive traits or shifts in the generalist community. The study by Guo et al. (2011) nicely demonstrates how important it is to consider more than one defence trait to gain a more complete picture about selection pressures and resource investment in plants of invasive populations.

In several studies, no significant differences in total defence concentrations between ranges were found but large differences were detected between populations (Müller and Martens, 2005; Huberty et al., 2014). Such differences may be imposed by local variation in herbivore communities (see also geographic mosaic; Thompson, 1999) and (allelopathic) interactions with competitors, causing location-specific evolutionary dynamics (Müller-Schärer et al., 2004). A particularly high qualitative variation in defences can be found (even within populations) in species that form different chemotypes, which can be discriminated on the basis of the presence of specific distinct metabolites that are dominant in groups of individuals of this species (Kleine and Müller, 2011). Such high chemical diversity might impede the adaptation abilities of antagonists and thus potentially contribute to the invasion success of these plant species (Macel et al., 2004; Kleine and Müller, 2011; Wolf et al., 2011; Fortuna et al., 2014).

The number of studies on indirect defences investigated in common gardens (also considered in my recent literature survey) is still very low, and the three known studies all focused on one species, Triadica sebifera (Carrillo et al., 2012, 2014; Wang et al., 2013; Table 12.1). The number of leaves with extrafloral nectaries was significantly lower in invasive populations of this species, indicating that, in the absence of herbivores, plants invest less in these costly structures (Carrillo et al., 2012; Wang et al., 2013). Further research on indirect defence traits is needed to be able to draw some general conclusions.

Shortcomings of the Distinction Between 'Qualitative' and 'Quantitative' Defence

One potential problem with the SDH is the distinction between cheap and expensive defences or between toxins, digestibility reducers and mechanical defences. Only very few studies challenged the issue to calculate metabolic costs for these types of defences. For example, although glucosinolates are considered as cheap, costs for their production have been revealed (Agrawal et al., 1999; Siemens et al., 2002; Cipollini *et al.*, 2003). In a model of the metabolism of *Arabidopsis thaliana* (Brassicaceae) using flux balance analysis, the production of glucosinolates increased photosynthetic requirements by at least 15% (Bekaert *et al.*, 2012). Thus, can these metabolites indeed be considered cheap? The association of terpenoids to either group of defences is even more problematic. The metabolic costs are considered to be high (and thus typical for quantitative defences) owing to their extensive chemical reduction, specific enzymes needed and expensive storage structures (Gershenzon, 1994). Yet, when comparing terpenoid concentrations of native and invasive populations of *Tanacetum vulgare*, higher terpenoid concentrations were found in invasive populations (Table 12.1), which is more characteristic of qualitative defences in terms of the SDH (Wolf *et al.*, 2011). In line with this, terpenoids of *T. vulgare* are used for host plant recognition by specialist herbivores, which obviously have adapted to these compounds (Müller and Hilker, 2001). In general, carbon can be cheaply produced, whereas the specific biosynthetic pathway for defence compounds based on carbon may be rather expensive.

The classification into qualitative and quantitative defences may also depend on the mode-of-action as well as the context. For example, iridoid glycosides, which consist of a monoterpene and a sugar moiety (Bowers, 1991), could be considered as typical qualitative defences because they act as a deterrent to various non-adapted insects (Dobler *et al.*, 2011). However, iridoid glycosides present a dual defence together with β-glucosidases (Pankoke *et al.*, 2013), leading to the production of highly reactive aglycones when coming into contact; the aglycones react with nucleophilic side chains of amino acids, forming covalent protein complexes (Konno *et al.*, 1999; Dobler *et al.*, 2011). In that way, the activation of iridoid glycosides by the enzymes causes digestibility-reducing effects associated with quantitative defences. Thus, a clear distinction between qualitative and quantitative defences might not always be possible and, instead, these defence types might form a continuum.

Various environmental factors have a huge impact on resource allocation in defence traits. Both the EICA hypothesis and the SDH are based on the assumption that resources are scarce and therefore should be allocated in growth or defence, depending on the needs or selection pressures. The costs for certain defences, however, depend on the environment in which a plant is growing (Doorduin and Vrieling, 2011). For example, toughening of leaves, considered as mechanical defence, incurs no costs in a sunny environment (Doorduin and Vrieling, 2011). Furthermore, changes in traits cannot only be interpreted in the light of shifting herbivore pressures. For example, trichomes serve as herbivore defence but also play important roles in temperature regulation, light reflection and transpiration (Smith and Nobel, 1977; Brewer *et al.*, 1991). The various examples highlight that care should be taken when dividing defences into cheap vs expensive or qualitative vs quantitative.

Conclusions and Future Directions

The EICA hypothesis is still tested in recent studies but, as reviewed earlier (Hinz and Schwarzlaender, 2004; Bossdorf *et al.*, 2005; Orians and Ward, 2010) and recently (Fig. 12.2, Table 12.1), it can only rarely explain invasion success and relatively few studies were supportive for the EICA hypothesis. A recent meta-analysis on the EICA hypothesis provides evidence that herbivore release does not generally act as a selection pressure on plant allocation into defence vs growth, but instead stochastic (e.g. founder events, multiple introductions, hybridization, bottlenecks or isolation by distance; Keller and Taylor, 2008) or selective forces other than herbivores (e.g. climate, resource availability) might be more important (Felker-Quinn *et al.*, 2013). Thus, the evolutionary divergence may be stronger between introduced-range plants and parental-range plants in the course of plant invasion (Felker-Quinn *et al.*, 2013). In contrast to the EICA hypothesis, the SDH distinguishes between different types of defences of distinct costs, which might, however, rather form a continuum. Researchers need to get a step further now and consider multiple levels of defences within one system, ideally testing plants in common-garden studies under more natural situations in which plants are exposed to competition as well as different environmental conditions (Kleine *et al.*, 2017). Isotope labelling and theoretical modelling might help to understand allocation patterns of, for example, carbon and nitrogen in growth-related or resistance-related traits. Furthermore, the dynamics of such allocation patterns need to be considered (Siemann and Rogers, 2001). Long-term manipulative experiments using geographic contrasts (Uesugi and Kessler, 2016) may be a great source for novel insights. Finally, studying the variation in defence allocation patterns between and within populations of invasive species will have important implications for biocontrol.

References

Abhilasha, D. and Joshi, J. (2009) Enhanced fitness due to higher fecundity, increased defence against a specialist and tolerance towards a generalist herbivore in an invasive annual plant. *Journal of Plant Ecology* 2, 77–86.

Adler, F.R. (1999) The balance of terror: An alternative mechanism for competitive trade-offs and its implications for invading species. *American Naturalist* 154, 497–509.

Agrawal, A.A., Strauss, S.Y. and Stout, M.J. (1999) Costs of induced responses and tolerance to herbivory in male and female fitness components of wild radish. *Evolution* 53, 1093–1104.

Agrawal, A.A., Hastings, A.P., Bradburd, G.S., Woods, E.C., Züst, T., Harvey, J.A. and Bukovinszky, T. (2015) Evolution of plant growth and defense in a continental introduction. *American Naturalist* 186, E1–E15.

Alba, C., Bowers, M.D. and Hufbauer, R. (2012) Combining optimal defense theory and the evolutionary dilemma model to refine predictions regarding plant invasion. *Ecology* 93, 1912–1921.

Bazzaz, F.A., Chiariello, N.R., Coley, P.D. and Pitelka, L.F. (1987) Allocating resources to reproduction and defense. *BioScience* 87, 58–67.

Bekaert, M., Edger, P.P., Hudson, C.M., Pires, J.C. and Conant, G.C. (2012) Metabolic and evolutionary costs of herbivory defense: systems biology of glucosinolate synthesis. *New Phytologist* 196, 596–605.

Blair, A.C. and Wolfe, L.M. (2004) The evolution of an invasive plant: An experimental study with *Silene latifolia*. *Ecology* 85, 3035–3042.

Blossey, B. and Nötzold, R. (1995) Evolution of increased competitive ability in invasive non-indigenous plants: a hypothesis. *Journal of Ecology* 83, 887–889.

Bossdorf, O., Prati, D., Auge, H. and Schmid, B. (2004) Reduced competitive ability in an invasive plant. *Ecology Letters* 7, 346–353.

Bossdorf, O., Auge, H., Lafuma, L., Rogers, W.E., Siemann, E. and Prati, D. (2005) Phenotypic and genetic differentiation between native and introduced plant populations. *Oecologia* 144, 1–11.

Bossdorf, O., Richards, C.L. and Pigliucci, M. (2008) Epigenetics for ecologists. *Ecology Letters* 11, 106–115.

Bowers, M.D. (1991) Iridoid glycosides. In: Rosenthal, G.A. and Berenbaum, M.R. (eds) *Herbivores: their Interactions with Secondary Plant Metabolites*, 2nd edn. Academic Press, New York, pp. 297–325.

Brewer, C.A., Smith, W.K. and Vogelmann, T.C. (1991) Functional interaction between leaf trichomes, leaf wettability and optical properties of water droplets. *Plant Cell and Environment* 14, 955–962.

Buschmann, H., Edwards, P.J. and Dietz, H. (2005) Variation in growth pattern and response to slug damage among native and invasive provenances of four perennial Brassicaceae species. *Journal of Ecology* 93, 322–334.

Callaway, R.M., Thelen, G.C., Rodriguez, A. and Holben, W.E. (2004) Soil biota and exotic plant invasion. *Nature* 427, 731–733.

Campbell, L.G., Parker, R.J., Blakelock, G., Pirimova, N. and Mercer, K.L. (2015) Maternal environment influences propagule pressure of an invasive plant, *Raphanus raphanistrum* (Brassicaceae). *International Journal of Plant Sciences* 176, 393–403.

Caño, L., Escarré, J., Vrieling, K. and Sans, F.X. (2009) Palatability to a generalist herbivore, defence and growth of invasive and native *Senecio* species: testing the evolution of increased competitive ability hypothesis. *Oecologia* 159, 95–106.

Carrillo, J., Wang, Y., Ding, J.Q., Klootwyk, K. and Siemann, E. (2012) Decreased indirect defense in the invasive tree, *Triadica sebifera*. *Plant Ecology* 213, 945–954.

Carrillo, J., McDermott, D. and Siemann, E. (2014) Loss of specificity: native but not invasive populations of *Triadica sebifera* vary in tolerance to different herbivores. *Oecologia* 174, 863–871.

Chew, F.S. and Courtney, S.P. (1991) Plant apparency and evolutionary escape from insect herbivory. *American Naturalist* 138, 729–750.

Cipollini, D. (2005) Interactive effects of lateral shading and jasmonic acid on morphology, phenology, seed production, and defense traits in *Arabidopsis thaliana*. *International Journal of Plant Sciences* 166, 955–959.

Cipollini, D., Purrington, C.B. and Bergelson, J. (2003) Costs of induced responses in plants. *Basic and Applied Ecology* 4, 79–85.

Cipollini, D.F., Mbagwu, J., Barto, K., Hillstrom, C. and Enright, S. (2005) Expression of constitutive and inducible chemical defenses in native and invasive populations of *Alliaria petiolata*. *Journal of Chemical Ecology* 31, 1255–1267.

Colautti, R.I., Maron, J.L. and Barrett, S.C.H. (2009) Common garden comparisons of native and introduced plant populations: latitudinal clines can obscure evolutionary inferences. *Evolutionary Applications* 2, 187–199.

Coley, P.D., Bryant, J.P. and Chapin, F.S. (1985) Resource availability and plant herbivore defense. *Science* 230, 895–899.

Daehler, C.C. and Strong, D.R. (1997) Reduced herbivore resistance in introduced smooth cordgrass (*Spartina alterniflora*) after a century of herbivore-free growth. *Oecologia* 110, 99–108.

Dietz, H. and Edwards, P.J. (2006) Recognition that causal processes change during plant invasion helps explain conflicts in evidence. *Ecology* 87, 1359–1367.

Dobler, S., Petschenka, G. and Pankoke, H. (2011) Coping with toxic plant compounds – the insect's perspective on iridoid glycosides and cardenolides. *Phytochemistry* 72, 1593–1604.

Doorduin, L.J. and Vrieling, K. (2011) A review of the phytochemical support for the shifting defence hypothesis. *Phytochemistry Reviews* 10, 99–106.

Eigenbrode, S.D., Andreas, J.E., Cripps, M.G., Ding, H., Biggam, R.C. and Schwarzlander, M. (2008) Induced chemical defenses in invasive plants: a case study with *Cynoglossum officinale* L. *Biological Invasions* 10, 1373–1379.

Engelkes, T., Morriën, E., Verhoeven, K.J.F., Bezemer, T.M., Biere, A., Harvey, J.A., McIntyre, L.M., Tamis, W.L.M and van der Putten, W.H. (2008) Successful range-expanding plants experience less above-ground and below-ground enemy impact. *Nature* 456, 946–948.

Felker-Quinn, E., Bailey, J.K. and Schweitzer, J.A. (2011) Soil biota drive expression of genetic variation and development of population-specific feedbacks in an invasive plant. *Ecology* 92, 1208–1214.

Felker-Quinn, E., Schweitzer, J.A. and Bailey, J.K. (2013) Meta-analysis reveals evolution in invasive plant species but little support for Evolution of Increased Competitive Ability (EICA). *Ecology and Evolution* 3, 739–751.

Feng, Y.L., Lei, Y.B., Wang, R.F., Callaway, R., Valiente-Banuet, A., Inderjit, Li, Y.P. and Zheng, Y.L. (2009) Evolutionary tradeoffs for nitrogen allocation to photosynthesis versus cell walls in an invasive plant. *Proceedings of the National Academy of Sciences USA* 106, 1853–1856.

Fortuna, T. M., Eckert, S., Harvey, J. A., Vet, L. E. M., Müller, C. and Gols, R. (2014) Variation in plant defences among populations of a range-expanding plant: consequences for trophic interactions. *New Phytologist,* 204, 989–999.

Franks, S.J., Pratt, P.D., Dray, F.A. and Simms, E.L. (2008) No evolution of increased competitive ability or decreased allocation to defense in *Melaleuca quinquenervia* since release from natural enemies. *Biological Invasions* 10, 455–466.

Franks, S.J., Wheeler, G.S. and Goodnight, C. (2012) Genetic variation and evolution of secondary compounds in native and introduced populations of the invasive plant *Melaleuca quinquenervia*. *Evolution* 66, 1398–1412.

Gao, L., Geng, Y., Li, B., Chen, J. and Yang, J. (2010) Genome-wide DNA methylation alterations of *Alternanthera philoxeroides* in natural and manipulated habitats: implications for epigenetic regulation of rapid responses to environmental fluctuation and phenotypic variation. *Plant Cell and Environment* 33, 1820–1827.

Gershenzon, J. (1994) Metabolic costs of terpenoid accumulation in higher plants. *Journal of Chemical Ecology* 20, 1281–1328.

Gillies, S., Clements, D.R. and Grenz, J. (2016) Knotweed (*Fallopia* spp.) invasion of North America utilizes hybridization, epigenetics, seed dispersal (unexpectedly), and an arsenal of physiological tactics. *Invasive Plant Science and Management* 9, 71–80.

Guo, W.F., Zhang, J., Li, X.Q. and Ding, J.Q. (2011) Increased reproductive capacity and physical defense but decreased tannin content in an invasive plant. *Insect Science* 18, 521–532.

Herms, D.A. and Mattson, W.J. (1992) The dilemma of plants – to grow or defend. *Quarterly Review of Biology* 67, 283–335.

Hinz, H.L. and Schwarzlaender, M. (2004) Comparing invasive plants from their native and exotic range: what can we learn for biological control. *Weed Technology* 18, 1533–1541.

Hornoy, B., Atlan, A., Tarayre, M., Dugravot, S. and Wink, M. (2012) Alkaloid concentration of the invasive plant species *Ulex europaeus* in relation to geographic origin and herbivory. *Naturwissenschaften* 99, 883–892.

Huang, W., Siemann, E., Wheeler, G.S., Zou, J.W., Carrillo, J. and Ding, J.Q. (2010) Resource allocation to defence and growth are driven by different responses to generalist and specialist herbivory in an invasive plant. *Journal of Ecology* 98, 1157–1167.

Huberty, M., Tielbörger, K., Harvey, J.A., Müller, C. and Macel, M. (2014) Chemical defenses of native and invasive populations of the range expanding invasive plant *Rorippa austriaca*. *Journal of Chemical Ecology* 40, 363–370.

Hull-Sanders, H.M., Clare, R., Johnson, R.H. and Meyer, G.A. (2007) Evaluation of the evolution of increased competitive ability (EICA) hypothesis: Loss of defense against generalist but not specialist herbivores. *Journal of Chemical Ecology* 33, 781–799.

Iacarella, J.C., Mankiewicz, P.S. and Ricciardi, A. (2015) Negative competitive effects of invasive plants change with time since invasion. *Ecosphere* 6, 123.

Johnson, R.H., Hull-Sanders, H.M. and Meyer, G.A. (2007) Comparison of foliar terpenes between native and invasive *Solidago gigantea*. *Biochemical Systematics and Ecology* 35, 821–830.

Joshi, J. and Vrieling, K. (2005) The enemy release and EICA hypothesis revisited: incorporating the fundamental difference between specialist and generalist herbivores. *Ecology Letters* 8, 704–714.

Joshi, S., Gruntman, M., Bilton, M., Seifan, M. and Tielbörger, K. (2014) A comprehensive test of evolutionarily increased competitive ability in a highly invasive plant species. *Annals of Botany* 114, 1761–1768.

Keane, R.M. and Crawley, M.J. (2002) Exotic plant invasions and the enemy release hypothesis. *Trends in Ecology & Evolution* 17, 164–170.

Keller, S.R. and Taylor, D.R. (2008) History, chance and adaptation during biological invasion: separating stochastic phenotypic evolution from response to selection. *Ecology Letters* 11, 852–866.

Kleine, S. and Müller, C. (2011) Intraspecific plant chemical diversity and its effects on herbivores. *Oecologia* 166, 175–186.

Kleine, S., Weißinger, L. and Müller, C. (2017) Impact of drought on plant populations of native and invasive origins. *Oecologia* 183, 9–20.

Konno, K., Hirayama, C., Yasui, H. and Nakamura, M. (1999) Enzymatic activation of oleuropein: A protein crosslinker used as a chemical defense in the privet tree. *Proceedings of the National Academy of Sciences USA* 96, 9159–9164.

Koricheva, J., Nykanen, H. and Gianoli, E. (2004) Meta-analysis of trade-offs among plant anti-herbivore defenses: Are plants jacks-of-all-trades, masters of all? *American Naturalist* 163, E64–E75.

Lankau, R.A. (2011) Intraspecific variation in allelochemistry determines an invasive species' impact on soil microbial communities. *Oecologia* 165, 453–463.

Lei, Y.B., Feng, Y.L., Zheng, Y.L., Wang, R.F., Gong, H.D. and Zhang, Y.P. (2011) Innate and evolutionarily increased advantages of invasive *Eupatorium adenophorum* over native *E. japonicum* under ambient and doubled atmospheric CO_2 concentrations. *Biological Invasions* 13, 2703–2714.

Liao, Z.Y., Zhang, R., Barclay, G.F. and Feng, Y.L. (2013) Differences in competitive ability between plants from nonnative and native populations of a tropical invader relates to adaptive responses in abiotic and biotic environments. *PloS ONE* 8, e71767.

Liao, Z.Y., Zheng, Y.L., Lei, Y.B. and Feng, Y.L. (2014) Evolutionary increases in defense during a biological invasion. *Oecologia* 174, 1205–1214.

Lin, T.T., Doorduin, L., Temme, A., Pons, T.L., Lamers, G.E.M., Anten, N.P.R. and Vrieling, K. (2015) Enemies lost: parallel evolution in structural defense and tolerance to herbivory of invasive *Jacobaea vulgaris*. *Biological Invasions* 17, 2339–2355.

Macel, M., de Vos, R. C.H., Jansen, J.J., van der Putten, W.H. and van Dam, N.M. (2014) Novel chemistry of invasive plants: exotic species have more unique metabolomic profiles than native congeners. *Ecology and Evolution* 4, 2777–2786.

Macel, M., Vrieling, K. and Klinkhamer, P.G.L. (2004) Variation in pyrrolizidine alkaloid patterns of *Senecio jacobaea*. *Phytochemistry* 65, 865–873.

Maron, J.L., Vilá, M. and Arnason, J. (2004a) Loss of enemy resistance among introduced populations of St. John's Wort (*Hypericum perforatum*). *Ecology* 85, 3243–3253.

Maron, J.L., Vilà, M., Bommarco, R., Elmendorf, S. and Beardsley, P. (2004b) Rapid evolution of an invasive plant. *Ecological Monographs* 74, 261–280.

McKey, D. (1979) The distribution of secondary plant compounds within plants. In: Rosenthal,

G.A. and Berenbaum, M.R. (eds) *Herbivores: their Interactions with Secondary Plant Metabolites*, 2nd edn. Academic Press, New York, pp. 55–133.

Mitchell, C.E. and Power, A.G. (2003) Release of invasive plants from fungal and viral pathogens. *Nature* 421, 625–627.

Mithen, R., Fraybould, A.F. and Giamoustaris, A. (1995) Divergent selection for secondary metabolites between wild populations of *Brassica oleracea* and its implications for plant-herbivore interactions. *Heredity* 75, 472–484.

Monty, A., Lebeau, J., Meerts, P. and Mahy, G. (2009) An explicit test for the contribution of environmental maternal effects to rapid clinal differentiation in an invasive plant. *Journal of Evolutionary Biology* 22, 917–926.

Morriën, E., Engelkes, T., Macel, M., Meisner, A. and van der Putten, W.H. (2010) Climate change and invasion by intracontinental range-expanding exotic plants: the role of biotic interactions. *Annals of Botany* 105, 843–848.

Müller, C. (2009) Role of glucosinolates in plant invasiveness. *Phytochemistry Reviews* 8, 227–242.

Müller, C. and Hilker, M. (2001) Host finding and oviposition behavior in a chrysomelid specialist – the importance of host plant surface waxes. *Journal of Chemical Ecology* 27, 985–994.

Müller, C. and Martens, N. (2005) Testing predictions of the 'evolution of increased competitive ability' hypothesis for an invasive crucifer. *Evolutionary Ecology* 19, 533–550.

Müller-Schärer, H. and Steinger, T. (2004) Predicting evolutionary change in invasive, exotic plants and its consequence for plant-herbivore interactions. In: Ehler, L.E., Sforza, R. and Mateille, T. (eds) *Genetics, Evolution and Biological Control*. CAB International, Wallingford, UK, pp. 137–162.

Müller-Schärer, H., Schaffner, U. and Steinger, T. (2004) Evolution in invasive plants: implications for biological control. *Trends in Ecology & Evolution* 19, 417–422.

Noble, I.R. (1989) Attributes of invaders and the invading process: terrestrial and vascular plants. In: Drake, J.A., Mooney, H.A., di Castri, F., Groves, R.H., Kruger, F.J., Reimánek, M. and Williamson, M. (eds) *Biological Invasions: a Global Perspective* – SCOPE 37. Wiley, Chichester, UK, pp. 301–313.

Oduor, A.M.O., Lankau, R.A., Strauss, S.Y. and Gómez, J.M. (2011) Introduced *Brassica nigra* populations exhibit greater growth and herbivore resistance but less tolerance than native populations in the native range. *New Phytologist* 191, 536–544.

Oduor, A.M.O., Stift, M. and van Kleunen, M. (2015) The interaction between root herbivory and competitive ability of native and invasive-range populations of *Brassica nigra*. *PloS ONE* 10, e0141857.

Orians, C.M. and Ward, D. (2010) Evolution of plant defenses in nonindigenous environments. *Annual Review of Entomology* 55, 439–459.

Pankoke, H., Buschmann, T. and Müller, C. (2013) Role of plant beta-glucosidases in the dual defense system of iridoid glycosides and their hydrolyzing enzymes in *Plantago lanceolata* and *Plantago major*. *Phytochemistry* 94, 99–107.

Pregitzer, C.C., Bailey, J.K., Hart, S.C. and Schweitzer, J.A. (2010) Soils as agents of selection: feedbacks between plants and soils alter seedling survival and performance. *Evolutionary Ecology* 24, 1045–1059.

Qing, H., Cai, Y., Xiao, Y., Yao, Y.H. and An, S.Q. (2012) Leaf nitrogen partition between photosynthesis and structural defense in invasive and native tall form *Spartina alterniflora* populations: effects of nitrogen treatments. *Biological Invasions* 14, 2039–2048.

Rapo, C., Müller-Scharer, H., Vrieling, K. and Schaffner, U. (2010) Is there rapid evolutionary response in introduced populations of tansy ragwort, *Jacobaea vulgaris*, when exposed to biological control? *Evolutionary Ecology* 24, 1081–1099.

Rice, K.J., Gerlach, J.D., Dyer, A.R. and McKay, J.K. (2013) Evolutionary ecology along invasion fronts of the annual grass *Aegilops triuncialis*. *Biological Invasions* 15, 2531–2545.

Richards, C.L., Schrey, A.W. and Pigliucci, M. (2012) Invasion of diverse habitats by few Japanese knotweed genotypes is correlated with epigenetic differentiation. *Ecology Letters* 15, 1016–1025.

Ridenour, W.M., Vivanco, J.M., Feng, Y.L., Horiuchi, J. and Callaway, R.M. (2008) No evidence for trade-offs: *Centaurea* plants from America are better competitors and defenders. *Ecological Monographs* 78, 369–386.

Roach, D.A. and Wulff, R.D. (1987) Maternal effects in plants. *Annual Review of Ecology and Systematics* 18, 209–235.

Schuman, M.C. and Baldwin, I.T. (2016) The layers of plant responses to insect herbivores. *Annual Review of Entomology* 61, 373–394.

Schweiger, R. and Müller, C. (2015) Leaf metabolome in arbuscular myccorhizal symbiosis. *Current Opinion in Plant Biology* 26, 120–126.

Schweitzer, J.A., Juric, I., van de Voorde, T.F.J., Clay, K., van der Putten, W.H. and Bailey, J.K. (2014) Are there evolutionary consequences of plant-soil feedbacks along soil gradients? *Functional Ecology* 28, 55–64.

Siemann, E. and Rogers, W.E. (2001) Genetic differences in growth of an invasive tree species. *Ecology Letters* 4, 514–518.

Siemann, E. and Rogers, W.E. (2003) Increased competitive ability of an invasive tree may be limited by an invasive beetle. *Ecological Applications* 13, 1503–1507.

Siemens, D.H., Garner, S.H., Mitchell-Olds, T. and Callaway, R.M. (2002) Cost of defense in the context of plant competition: *Brassica rapa* may grow *and* defend. *Ecology* 83, 505–517.

Smith, W.K. and Nobel, P.S. (1977) Influences of seasonal changes in leaf morphology on water-use efficiency for three desert broadleaf shrubs. *Ecology* 58, 1033–1043.

Stastny, M., Schaffner, U. and Elle, E. (2005) Do vigour of introduced populations and escape from specialist herbivores contribute to invasiveness? *Journal of Ecology* 93, 27–37.

Strauss, S.Y., Irwin, R.E. and Lambrix, V.M. (2004) Optimal defence theory and flower petal colour predict variation in the secondary chemistry of wild radish. *Journal of Ecology* 92, 132–141.

Tewes, L.J., Michling, F., Koch, M.A. and Müller C. (2018) Chemical diversity mirrors genetic diversity and might represent a key advantage in intracontinental invasion of the perennial *Bunias orientalis*. *Journal of Ecology*. DOI: 10.1111/1365-2745.12869

Thompson, J.N. (1999) Specific hypotheses on the geographic mosaic of coevolution. *American Naturalist* 153, S1–S14.

Travers-Martin, N. and Müller, C. (2008) Matching plant defense syndromes with performance and preference of a specialist herbivore. *Functional Ecology* 22, 1033–1043.

Uesugi, A. and Kessler, A. (2016) Herbivore release drives parallel patterns of evolutionary divergence in invasive plant phenotypes. *Journal of Ecology* 104, 876–886.

van der Meijden, E. (1996) Plant defence, an evolutionary dilemma: contrasting effects of (specialist and generalist) herbivores and natural enemies. *Entomologia Experimentalis et Applicata* 80, 307–310.

van Kleunen, M. and Schmid, B. (2003) No evidence for an evolutionary increased competitive ability in an invasive plant. *Ecology* 84, 2816–2823.

Wang, Y., Siemann, E., Wheeler, G.S., Zhu, L., Gu, X. and Ding, J.Q. (2012) Genetic variation in anti-herbivore chemical defences in an invasive plant. *Journal of Ecology* 100, 894–904.

Wang, Y., Carrillo, J., Siemann, E., Wheeler, G.S., Zhu, L., Gu, X. and Ding, J. (2013) Specificity of extrafloral nectar induction by herbivores differs among native and invasive populations of tallow tree. *Annals of Botany* 112, 751–756.

Willis, A.J., Thomas, M.B. and Lawton, J.H. (1999) Is the increased vigour of invasive weeds explained by a trade-off between growth and herbivore resistance? *Oecologia* 120, 632–640.

Wolf, V.C., Berger, U., Gassmann, A. and Müller, C. (2011) High chemical diversity of a plant species is accompanied by increased chemical defence in invasive populations. *Biological Invasions* 13, 2091–2102.

Zheng, Y.L., Feng, Y.L., Liao, Z.Y., Li, W.T., Xiao, H.F. and Sui, H.Z. (2013) Invasive *Chromolaena odorata* has similar size but higher phenolic concentration than native conspecifics. *Evolutionary Ecology Research* 15, 769–781.

Zheng, Y.L., Feng, Y.L., Zhang, L.K., Callaway, R.M., Valiente-Banuet, A., Luo, D.-Q., Liao, Z.-Y., Lei, Y.-B., Barclay, G.F. and Silva-Pereyra, C. (2015) Integrating novel chemical weapons and evolutionarily increased competitive ability in success of a tropical invader. *New Phytologist* 205, 1350–1359.

13 Tens Rule

Jonathan M. Jeschke[1,2,3]* and Petr Pyšek[4,5]

[1]Leibniz-Institute of Freshwater Ecology and Inland Fisheries
(IGB), Berlin, Germany; [2]Freie Universität Berlin, Institute of
Biology, Berlin, Germany; [3]Berlin-Brandenburg Institute of
Advanced Biodiversity Research (BBIB), Berlin, Germany;
[4]Institute of Botany, Department of Invasion Ecology, The Czech
Academy of Sciences, Průhonice, Czech Republic; [5]Charles
University, Faculty of Science, Department of Ecology, Prague,
Czech Republic

Abstract

The tens rule became a popular invasion hypothesis in the 1990s and is still widely used today, even though empirical support has been mixed from the beginning and the number of studies questioning it has been increasing in the past decade. Also, the rule is not based on a model or other defensible concept or argument. Here we divide the tens rule into two more specific sub-hypotheses: the *invasion tens rule* and the *impact tens rule*, where the former predicts that about 10% of species successfully take consecutive steps of the invasion process, and the latter that about 10% of established non-native species and about 1% of all introduced non-native species cause significant detrimental impacts. A quantitative meta-analysis of 102 empirical tests of the tens rule from 65 publications shows no support for this hypothesis. Looking at the invasion tens rule and comparing different taxonomic groups, about 25% of non-native plants and invertebrates, and about 50% of non-native vertebrates are on average successful in taking consecutive steps of the invasion process. We thus suggest replacing the invasion tens rule by two taxon-dependent hypotheses: the *50% invasion rule* for vertebrates and the *25% invasion rule* for other organisms,

particularly plants and invertebrates. The impact tens rule is not supported by currently available evidence, either, and more data are needed before a reasonable alternative hypothesis can be formulated. In a nutshell, we suggest abandoning the tens rule and using the 50% invasion rule for vertebrates and the 25% invasion rule for other organisms. These hypotheses provide new standards that are supported by currently available data and against which future data can be tested.

Introduction

The tens rule posits that about 10% of species successfully take consecutive steps of the invasion process: circa 10% of species transported beyond their native range will be released or escape in the wild (they are called introduced species or casuals); about 10% of these introduced species will be able to establish viable populations in the wild (they are often called naturalized species); and about 10% of species established will become invasive/pest species (Williamson and Brown, 1986; Williamson, 1993, 1996; Jeschke *et al.*, 2012; Jeschke, 2014). This rule became popular in the late 20th century

---

* Corresponding author. E-mail: jonathan.jeschke@gmx.net

and has been very influential within the field of invasion biology and beyond. It can be found in many popularizations (e.g. Kegel, 2013) and exhibitions (e.g. in the botanic garden in Potsdam, Germany) and has also been applied to genetically modified organisms (Regal, 1993; Williamson, 1993). The probability of a species transiting through the invasion process is a highly important parameter for ecological–economic cost–benefit models on the usefulness of actions such as border controls, and these models are sensitive to the precision of this probability (Keller et al., 2007). It is thus not surprising that the tens rule has received considerable attention.

Yet this rule has several limitations and shortcomings. In particular, the proposed 10% value was not based on a model or other defensible concept or argument – it was simply picked on the basis of the idea that most non-native species will not be able to pass through the invasion process and have no significant impact. However, this idea was not further conceptualized or formally developed.

Another difficulty with this rule is that the invasion process to which it is linked has been differently defined by different authors. A particular challenge is that 'impact' is not really part of the invasion process because non-native species can have impacts during any stage of their invasion. Although their impact tends to increase through the invasion process, non-native species can have impacts directly after their arrival in the exotic range – think about a parasite or pathogen as an example (Ricciardi and Cohen, 2007; Ricciardi et al., 2013; Jeschke et al., 2013, 2014; Chapter 1, this volume). We thus follow Blackburn et al. (2011) and Jeschke et al. (2013) in not integrating 'impact' into the invasion process, but considering the following stages of this process: (i) *transport* to exotic range → (ii) *introduction* (release or escape into the environment) → (iii) *establishment* of a least one self-sustaining population → (iv) *spread*. These stages are as in Blackburn et al. (2011).

Because 'impact' is not a stage of the invasion process, we discriminate two different variants, i.e. sub-hypotheses, of the tens rule. The *invasion tens rule* (first sub-hypothesis) is restricted to the three outlined transitions between invasion stages. In addition, the suggestion that about 10% of established non-native species cause a significant detrimental impact (either on ecology/biodiversity, socioeconomics or human health) and that about 1% of all introduced species cause a significant detrimental impact can be termed the *impact tens rule* (second sub-hypothesis). We decided to devote attention to the impact tens rule in this chapter because it has been relatively unexplored thus far compared to the invasion tens rule (the latter can also be termed tens rule *sensu stricto*). Furthermore, Strayer (2012) suggested, on the basis of empirical evidence, that about 3–30% of established invaders substantially affect ecosystem functioning, which is in line with the impact tens rule. Please note that the suggestion that 1% of all introduced species cause a significant detrimental impact has not been made explicit in the context of the tens rule before.

A further difficulty with the tens rule is that its predictions are sensitive to the spatiotemporal scale of interest. Regarding the temporal scale, if more time passes, then typically more introduction events of a given species will have occurred for a given region, either intentionally or unintentionally, and some of these introductions will have been successful. One of the predictions of the tens rule is, as outlined above, that about 10% of all introduced species will establish themselves. When researchers tested this prediction for a given region and a given set of species, let's say in the 1990s, and another team of researchers repeats the study with the same set of species today, they will find a higher establishment success (defined as the number of established species divided by the number of introduced species) today than in the earlier study, assuming that no or few of the originally established species later died out. The tens rule predicts that establishment success is about 10% and does not qualify the temporal scale, yet the establishment success of a species is actually time dependent (see also Richardson and Pyšek, 2006).

Establishment success is also dependent on the spatial scale. For instance, let's again look at two teams of researchers that investigate establishment success. Team A has chosen the small European country Liechtenstein as their focal region, whereas team B has chosen the USA. If a given species has been introduced to both countries multiple times, there is a higher chance that it was able to find a suitable environment in the USA where a larger suite of environmental conditions are available as well as more species as potential positive interaction partners of the non-native species (e.g. prey, food or mutualists). In other words, there is a higher probability that the ecological niche of the non-native species fits somewhere in the USA compared to Liechtenstein. It is even more extreme when comparing global establishment success (of species introduced anywhere) with small-scale establishment success. To our knowledge, this limitation of the tens rule in its applicability across spatial scales has been largely overlooked thus far.

Given all these limitations, it might not surprise that the tens rule has received mixed empirical support at best (Jeschke et al., 2012; Jeschke, 2014 and references therein). Another complication when testing this hypothesis is that reliable data about establishment success are often hard to find because numbers of introduced species that did not establish are often unknown (failed introductions; Jeschke, 2009; Rodriguez-Cabal et al., 2009). They are typically best known for mammals and birds, which is why most studies addressing the tens rule have been done for vertebrates, in contrast to most other invasion hypotheses where the majority of studies focus on plants (Jeschke et al., 2012; Chapter 17, this volume).

Despite its limitations, the tens rule has remained a major hypothesis of the field and is still widely used today. For instance, in a recent survey among >350 experts by Enders et al. (2018), it was the seventh best-known out of 33 invasion hypotheses featured in the survey. In this chapter, we use a quantitative meta-analytic approach to address the following questions: (i) what is the level of support for the tens rule and its sub-hypotheses? (ii) Does the level of support differ among major taxonomic groups and habitats? (iii) Has the level of support changed over time?

Methods

Hierarchy of hypotheses

Using the hierarchy-of hypotheses (HoH) approach (Chapters 2 and 6, this volume), we divided the tens rule into the invasion tens rule and the impact tens rule as follows:

- Invasion tens rule: about 10% of species successfully take consecutive steps of the invasion process.
  i. Transport → introduction: about 10% of the transported non-native species are released or escape.
  ii. Introduction → establishment: about 10% of the introduced species are establishing themselves.
  iii. Establishment → spread: about 10% of the established species are substantially spreading from their point(s) of introduction.
- Impact tens rule:
  iv. About 10% of established non-native species cause a significant detrimental impact (i.e. they have harmful ecological, socio-economic or human health effects); this sub-hypothesis thus relates to the transition establishment → impact/pest species.
  v. About 1% of all introduced non-native species cause a significant detrimental impact; this sub-hypothesis thus relates to the transition introduction → impact/pest species.

Dataset

We updated a previously collected dataset (Jeschke et al., 2012) for our analyses. This dataset originated from a systematic literature search done in 2010 using the string '(tens rule OR establishment success) AND (alien OR exotic OR introduced OR invasive

OR naturali?ed OR nonindigenous OR non-native)'; see Jeschke *et al.* (2012) for details. The dataset includes 75 empirical tests of the tens rule from 53 publications (publications testing two or three transitions in the invasion process were considered as two or three tests of the tens rule). Most of these tests relate to the sub-hypothesis on the transition introduction → establishment (which is generally well investigated and was also emphasized in the search string).

Because the impact tens rule is not well represented in this previous dataset, we updated the dataset with an additional literature search focused on this hypothesis. We searched the Web of Science on 8 December 2016 using the following string: 'tens rule AND (impact* OR effect* OR affect* OR chang* OR ecosystem service* OR harm* OR pest*) AND (alien OR exotic OR introduced OR invasive OR naturali?ed OR nonindigenous OR non-native)'. In addition, we searched for publications in the Web of Science that cited Vilà *et al.* (2010), which is a key paper on the proportions of non-native species with impacts. Finally, one paper in Jeschke *et al.*'s. (2012) dataset was replaced: the 2nd edition of the catalogue of alien plants of the Czech Republic (Pyšek *et al.*, 2012) replaced the 1st edition (Pyšek *et al.*, 2002). The combined and updated dataset includes a total of 102 empirical tests from 65 publications. It is freely available online at www.hi-knowledge.org.

Quantitative meta-analysis

We applied a quantitative approach to compare the predictions of the tens rule with the data reported in the 102 tests we identified. For each sub-hypothesis, we calculated weighted means and 95% confidence intervals (CIs) for the percentage of species making the transition. We thereby followed a random-effects meta-analytic approach, using the DerSimonian–Laird method as implemented in the OpenMetaAnalyst software (Wallace *et al.*, 2012), which in turn uses the metafor package in R (Viechtbauer, 2010). After having calculated the means

and CIs in this way, we compared these to the 10% value predicted by the tens rule and the 5–20% range suggested by Williamson (1996); in the case of the transition introduction → impact/pest species, the comparison was done with the 1% prediction by the tens rule. We also compared the values across taxonomic groups, habitats and the time when empirical tests were published.

Results and Discussion

What is the level of support for the tens rule and its sub-hypotheses?

Neither the tens rule nor its sub-hypotheses are empirically supported by currently available evidence (Fig. 13.1). About two-thirds of the empirical tests in our dataset have focused on the invasion tens rule, the majority of these in turn on the transition introduction → establishment: about half of all empirical tests of the tens rule have focused on this sub-sub-hypothesis. The observed average percentage of species making this transition is >40% and thus more than four times larger than the tens rule's prediction; the difference is also statistically significant (Fig. 13.1). It is similar for the transition establishment → spread, where the observed percentage of species making the transition is >30% and thus more than three times larger than the prediction. The situation is less clear for the transition transport → introduction for which our dataset includes the lowest number of studies.

In the case of the impact tens rule, we observed that on average about 1 out of 4 established non-native species have a significant detrimental impact, which is again significantly more than the 1 out of 10 species predicted (Fig. 13.1). The discrepancy between observation and prediction is even larger for the transition introduction → impact/pest species: here, we observed that on average about 16 out of 100 alien species have a significant detrimental impact, whereas the impact tens rule predicts only 1 out of 100 alien species; hence there is a 16-fold difference here.

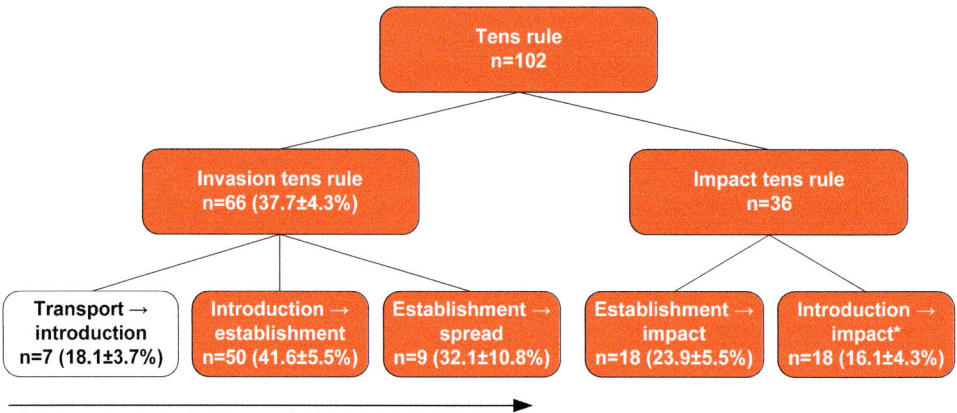

Consecutive invasion stages for invasion tens rule

Fig. 13.1. The hierarchy of hypotheses for the tens rule. The boxes are colour coded: red indicates that the observed percentage of species making the transition is questioning the tens rule, i.e. the mean is >20%, the 95% confidence interval (CI) does not overlap with 10% and n ≥5; green (not existent) would indicate that the percentage of species making the transition is in line with the tens rule, i.e. the mean is between 5 and 20%, the 95% CI overlaps with 10% and n ≥5; and white is used for other cases, i.e. inconclusive data or n <5. Detailed information on the percentage of species making each transition and 95% CIs are provided in parentheses. *For the transition introduction → impact/pest species, the tens rule predicts that only 1% of the species make this transition; the colour coding has been applied accordingly. Because of this basic difference of the introduction → impact rule to the other rules, no quantitative summary values are provided for the impact tens rule and overall tens rule. They are coloured in red because their sub-hypotheses are contradicted by the available empirical data.

These findings are in line with previous results based on smaller datasets (Jeschke et al., 2012; Jeschke, 2014 and references therein). Hence, currently available evidence does not support the tens rule. In the next section, we ask whether the tens rule is supported for some taxonomic groups or habitats, or whether this invasion hypothesis should be revised, replaced or completely abandoned.

Does the level of support differ among major taxonomic groups and habitats?

Even though the updated dataset has more data on the impact tens rule than the dataset of Jeschke et al. (2012), most data still focus on the invasion tens rule, and thus comparisons among taxonomic groups and habitats are particularly informative here. Our results show that much higher proportions of vertebrates than plants and

invertebrates are successful in taking consecutive steps of the invasion process: on average, about every fourth plant (24%) and invertebrate (23%) is successful, whereas every second vertebrate (51%) succeeds (Fig. 13.2a). This difference is statistically significant, and the transition probabilities of all three taxonomic groups are significantly higher than the 10% predicted by the tens rule (Fig. 13.2a). These findings are largely in line with Jeschke et al. (2012) who also found significantly lower support for the invasion tens rule for vertebrates than for plants and invertebrates.

Regarding the impact tens rule, on average 18% of established plants have shown detrimental impacts, which is still significantly higher than 10% but much closer to the tens rule's prediction than the average values for invertebrates and vertebrates, which are both above 30% (Fig. 13.2b). Sample sizes are low, however, for the impact tens rule, hence more studies are needed to test whether these values hold true. This is

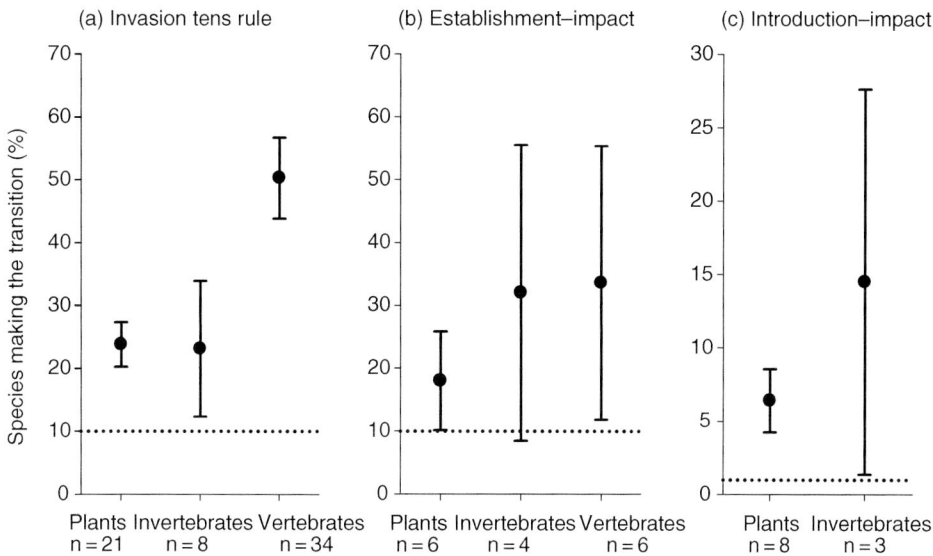

Fig. 13.2.   Differences among major taxonomic groups in the percentage of species making the transitions (a) between consecutive stages of the invasion process (invasion tens rule, all three transitions combined), (b) establishment → impact/pest species and (c) introduction → impact/pest species (vertebrate studies are not shown here owing to the very low sample size of n=2; panels b and c relate to the impact tens rule). Shown are means±95% confidence intervals. Sample sizes do not add up to 102 because empirical tests covering multiple taxonomic groups are not included. 'Plants' also include algae. Predicted percentages are indicated by dotted lines.

also true for the transition introduction → impact/pest species where data for vertebrates are very rare and thus not shown in Fig. 13.2c. Comparing plants and invertebrates, plants are again closer to the prediction of the tens rule than invertebrates. 'Closer' is relative, though, as on average 6% of the introduced plants have detrimental impacts, which is six times higher than the prediction; and 15% of the introduced invertebrates have detrimental impacts, which is 15 times higher than the prediction. These differences to the predicted value are also statistically significant (Fig. 13.2c).

The differences among major types of habitat (terrestrial, freshwater, marine) were slightly less pronounced (Fig. 13.3). Still, freshwater species were significantly more successful than terrestrial species in taking consecutive steps of the invasion process, with marine species being in between (invasion tens rule; Fig. 13.3a). This result is largely in line with Jeschke *et al.* (2012) who also found a significant difference between

terrestrial and freshwater species. Regarding the impact tens rule, observed values were again consistently higher than predicted values, in most cases significantly so (Fig. 13.3b,c).

Has the level of support changed over time?

A decline in the level of support for six invasion hypotheses was reported by Jeschke *et al.* (2012), and decline effects have been previously reported from other disciplines, particularly medicine, psychology and evolutionary ecology (Lehrer, 2010; Schooler, 2011). Underlying reasons include publication biases, biases in study organisms or systems and the psychology of researchers (Jeschke *et al.*, 2012, and references therein).

Our quantitative analysis on a possible decline effect did not include the transition introduction → impact/pest species because

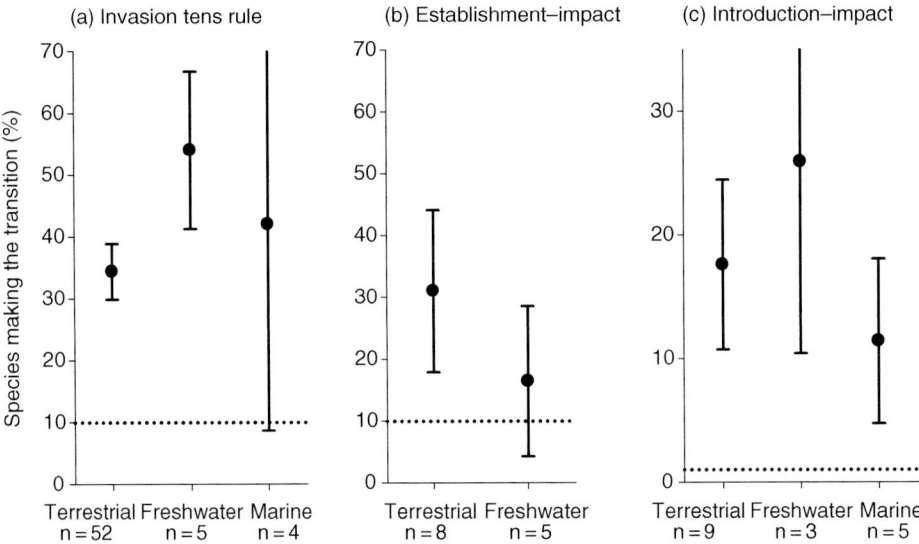

Fig. 13.3. Differences among major habitats in the percentage of species making the transitions (a) between consecutive stages of the invasion process (invasion tens rule, all three transitions combined), (b) establishment → impact/pest species (marine studies are not shown here due to the very low sample size of n=2) and (c) introduction → impact/pest species (panels b and c relate to the impact tens rule). Shown are means±95% confidence intervals. Sample sizes do not add up to 102, as empirical tests covering multiple habitats are not included. Predicted percentages are indicated by dotted lines.

the prediction for this transition is 1%; thus data cannot be pooled with data for the other transitions where the prediction is 10%. The remaining data were used for this analysis. These 84 studies tended to show increasing transition success rates over time (Pearson's correlation coefficient $r = 0.37$, $p = 0.099$). Thus over time, observed transition success rates tended to increasingly differ from the 10% value predicted by the tens rule. One could interpret this tendency as a slight decline effect that was, however, not statistically significant.

Conclusions

In a nutshell, the tens rule lacks empirical support and neither of its sub-hypotheses depicted in Fig. 13.1 – the invasion tens rule and impact tens rule – are supported by currently available evidence. This is in line with previous studies based on smaller datasets (Jeschke *et al.*, 2012; Jeschke, 2014, and references therein). Nonetheless, the rule has

remained a major hypothesis of the field and is still widely used today.

Jeremy Fox (2011), based on Quiggin (2010), used the term *zombie ideas* for hypotheses that are neither dead nor alive:

> Ideas, especially if they are widely believed, are intuitively appealing, and lack equally-intuitive replacements, tend to persist. And they persist not just in spite of a single inconvenient fact, but in spite of repeated theoretical refutations and whole piles of contrary facts. They are not truly alive— because they are not true—but neither are they dead. They are undead. They are zombie ideas.

It seems to us that the tens rule is a sort of zombie idea. On the basis of the findings reported here, we suggest that the invasion tens rule is replaced by two taxon-dependent hypotheses: the *50% invasion rule* for vertebrates and the *25% invasion rule* for other organisms, particularly plants and invertebrates. It should be kept in mind that these hypotheses share weaknesses with the tens rule, which are outlined in the Introduction above. One could argue that they should not be called 'rules' because of these weaknesses; however, we use this term, at least for now, in order to highlight that they are revisions of the tens *rule*. Like the tens rule but in contrast to other focal hypotheses in this book, they are data-driven and lack a formal theoretical foundation. But, in contrast to the tens rule, they are supported by currently available evidence.

Regarding the impact tens rule, more data are needed before it can be reasonably replaced by another hypothesis. The evidence that is currently available is too thin and the definitions of impact applied in available studies too variable (cf. Jeschke *et al.*, 2014) for an alternative hypothesis to be reasonably formulated at the moment. In any case, the percentage of introduced or established species with impact is not always the most important information because a single non-native species can have devastating impacts by itself. For example, an invasive lineage of the chytrid *Batrachochytrium dendrobatidis* is threatening a large number of amphibians worldwide, and rats, cats and other mammals are similarly threatening numerous vertebrate species (Bellard *et al.*, 2016). There is currently a lot of work focusing on invader impacts (reviewed in e.g. Jeschke *et al.*, 2014) and applications of the new IUCN Environmental Impact Classification for Alien Taxa (EICAT; Blackburn *et al.*, 2014; Hawkins *et al.*, 2015) might also prove useful for replacing the impact tens rule with a more adequate hypothesis.

References

Bellard, C., Genovesi, P. and Jeschke, J.M. (2016) Global patterns in threats to vertebrates by biological invasions. *Proceedings of the Royal Society B* 283, 20152454.

Blackburn, T.M., Pyšek, P., Bacher, S., Carlton, J.T., Duncan, R.P., Jarošík, V., Wilson, J.R.U. and Richardson, D.M. (2011) A proposed unified framework for biological invasions. *Trends in Ecology & Evolution* 26, 333–339.

Blackburn, T.M., Essl, F., Evans, T., Hulme, P.E., Jeschke, J.M., Kühn, I., Kumschick, S., Marková, Z., Mrugała, A., Nentwig, W. *et al.* (2014) A unified classification of alien species based on the magnitude of their environmental impacts. *PLoS Biology* 12, e1001850.

Enders, M., Hütt, M.-T. and Jeschke, J.M. (2018) Drawing a map of invasion biology based on a network of hypotheses. *Ecosphere.* DOI: 10.1002/ecs2.2146.

Fox, J. (2011) Zombie ideas in ecology. *Oikos blog.* Available at: oikosjournal.wordpress.com/2011/06/17/zombie-ideas-in-ecology (accessed 10 October 2017).

Hawkins, C.L., Bacher, S., Essl, F., Hulme, P.E., Jeschke, J.M., Kühn, I., Kumschick, S., Nentwig, W., Pergl, J., Pyšek, P. *et al.* (2015) Framework and guidelines for implementing the proposed IUCN Environmental Impact Classification for Alien Taxa (EICAT). *Diversity and Distributions* 21, 1360–1363.

Jeschke, J.M. (2009) Across islands and continents, mammals are more successful invaders than birds (Reply to Rodriguez-Cabal *et al.*). *Diversity and Distributions* 15, 913–914.

Jeschke, J.M. (2014) General hypotheses in invasion ecology. *Diversity and Distributions* 20, 1229–1234.

Jeschke, J.M., Gómez Aparicio, L., Haider, S., Heger, T., Lortie, C.J., Pyšek, P. and Strayer, D.L. (2012) Support for major hypotheses in invasion biology is uneven and declining. *NeoBiota* 14, 1–20.

Jeschke, J.M., Keesing, F. and Ostfeld, R.S. (2013) Novel organisms: comparing invasive species, GMOs, and emerging pathogens. *Ambio* 42, 541–548.

Jeschke, J.M., Bacher, S., Blackburn, T.M., Dick, J.T.A., Essl, F., Evans, T., Gaertner, M., Hulme, P. E., Kühn, I., Mrugała, A. *et al.* (2014) Defining the impact of non-native species. *Conservation Biology* 28, 1188–1194.

Kegel, B. (2013) *Die Ameise als Tramp: von Biologischen Invasionen.* 2nd edn. DuMont, Köln, Germany.

Keller, R.P., Lodge, D.M. and Finnoff, D.C. (2007) Risk assessment for invasive species produces net bioeconomic benefits. *Proceedings of the National Academy of Sciences USA* 104, 203–207.

Lehrer, J. (2010) The truth wears off. *New Yorker* Dec 13, 52–57.

Pyšek, P., Sádlo, J. and Mandák, B. (2002) Catalogue of alien plants of the Czech Republic. *Preslia* 74, 97–186.

Pyšek, P., Danihelka, J., Sádlo, J., Chrtek Jr., J., Chytrý, M. *et al.* (2012) Catalogue of alien plants of the Czech Republic (2nd edition): checklist update, taxonomic diversity and invasion patterns. *Preslia* 84, 155–255.

Quiggin, J. (2010) *Zombie Economics: How Dead Ideas Still Walk Among Us*. Princeton University Press, Princeton, New Jersey.

Regal, P.J. (1993) The true meaning of 'exotic species' as a model for genetically engineered organisms. *Experientia* 49, 225–234.

Ricciardi, A. and Cohen, J. (2007) The invasiveness of an introduced species does not predict its impact. *Biological Invasions* 9, 309–315.

Ricciardi, A., Hoopes, M.F., Marchetti, M.P. and Lockwood, J.L. (2013) Progress toward understanding the ecological impacts of nonnative species. *Ecological Monographs* 83, 263–282.

Richardson, D.M. and Pyšek, P. (2006) Plant invasions: merging the concepts of species invasiveness and community invasibility. *Progress in Physical Geography* 30, 409–431.

Rodriguez-Cabal, M.A., Barrios-Garcia, M.N. and Simberloff, D. (2009) Across islands and continents, mammals are more successful invaders than birds (Reply). *Diversity and Distributions* 15, 911–912.

Schooler, J. (2011) Unpublished results hide the decline effect. *Nature* 470, 437–437.

Strayer, D. L. (2012) Eight questions about invasions and ecosystem functioning. *Ecology Letters* 15, 1199–1210.

Viechtbauer, W. (2010) Conducting meta-analysis in R with the metafor package. *Journal of Statistical Software* 36, 3.

Vilà, M., Basnou, C., Pyšek, P., Josefsson, M., Genovesi, P., Gollasch, S., Nentwig, W., Olenin, S., Roques, A., Roy, D., Hulme, P. E. and DAISIE partners (2010) How well do we understand the impacts of alien species on ecosystem services? A pan-European, cross-taxa assessment. *Frontiers in Ecology and the Environment* 8, 135–144.

Wallace, B.C., Dahabreh, I.J., Trikalinos, T.A., Lau, J., Trow, P. and Schmid, C.H. (2012) Closing the gap between methodologists and end-users: R as a computational back-end. *Journal of Statistical Software* 49, 5.

Williamson, M. (1993) Invaders, weeds and the risk from genetically modified organisms. *Experientia* 49, 219–224.

Williamson, M. (1996) *Biological Invasions*. Chapman & Hall, London.

Williamson, M. and Brown, K.C. (1986) The analysis and modelling of British invasions. *Philosophical Transactions of the Royal Society B* 314, 505–522.

14 Phenotypic Plasticity Hypothesis

Olena Torchyk[1] and Jonathan M. Jeschke[1,2,3,4*]

[1]Technical University of Munich, Restoration Ecology, Freising, Germany; [2]Leibniz-Institute of Freshwater Ecology and Inland Fisheries (IGB), Berlin, Germany; [3]Freie Universität Berlin, Institute of Biology, Berlin, Germany; [4]Berlin-Brandenburg Institute of Advanced Biodiversity Research (BBIB), Berlin, Germany

Abstract

The phenotypic plasticity hypothesis – or, in short, plasticity hypothesis – posits that invasive species are more phenotypically plastic than non-invasive or native ones. On the basis of a systematic review, we identified 115 relevant empirical tests of the plasticity hypothesis. Most of these empirical studies focused on terrestrial plants; only some have been carried out on animals or in aquatic habitats. The plasticity hypothesis is largely empirically supported, including most of its sub-hypotheses – focusing on phenotypic plasticity in morphology, physiology and life history – and across taxonomic groups and habitats. There are relatively few experimental field and enclosure studies available, and these showed significantly lower support than observational studies. Similarly, recent studies showed slightly lower support of the plasticity hypothesis than early ones but this difference was not statistically significant. Thus overall, this invasion hypothesis is largely supported by currently available evidence; however, more studies are needed on organisms other than plants and in aquatic habitats, and it seems important to perform more experimental field and enclosure studies in the future to scrutinize this hypothesis.

Introduction

Phenotypic plasticity is the ability of a genotype to have variable phenotypes in different environments. It is a widespread phenomenon: rather than the presence or absence of phenotypic plasticity, it is the degree of phenotypic plasticity that differs among genotypes (Tollrian and Harvell, 1999; DeWitt and Scheiner, 2004; Engel et al., 2011; Utz et al., 2014). Well-known examples that have a high level of phenotypic plasticity include *Daphnia* waterfleas, which are able to build up neckteeth and similar morphological defences against invertebrate predators such as phantom midge larvae (*Chaoborus* spp.) when they live in water bodies with such predators; they do not build these defences in the absence of invertebrate predators and express other defences such as a more transparent and smaller body in order to reduce predation pressure from fishes (Tollrian and Harvell, 1999). Other well-known examples of phenotypic plasticity include acclimation, changes in chemical composition or organ size (e.g. many snakes, birds and lactating mammals can dramatically increase the size of their gut), or in life-history traits such as age of first reproduction or fecundity (for further details, see DeWitt and Scheiner, 2004; Utz et al., 2014, and references therein).

* Corresponding author. E-mail: jonathan.jeschke@gmx.net

Changing environments favour the evolution of phenotypic plasticity. Taking the example of inducible defences again, it is advantageous for a genotype to be able to build up different defences if it is confronted with changing predator regimes, particularly when predators differ in their hunting strategies and prey size preferences. For example, invertebrate predators such as *Chaoborus* spp. are tactile hunting predators that are more strongly limited in the prey size they can handle compared to fishes, which are also visual predators. A similar reasoning applies to other types of phenotypic plasticity. Under constant environmental conditions, phenotypic plasticity is often disadvantageous because being plastic typically also involves costs for organisms; for instance, resources are needed to build up detection mechanisms for environmental conditions. Conversely, if environmental conditions markedly change, then highly phenotypically plastic organisms typically have an advantage over other organisms because they can quickly adjust to such conditions.

Similarly, because non-native species arrive in ecosystems that often differ quite dramatically from their native ecosystems, being plastic allows them to adjust their phenotype quickly in order to increase their chances to survive, grow and reproduce. In addition, a higher level of phenotypic plasticity might also give non-native species a competitive advantage over other species (either native or other non-native ones) with lower levels of phenotypic plasticity that have been adapted to an environment that is not quickly changing. In the current era of the Anthropocene, environmental change has been accelerated because of human action and is predicted to accelerate further (Waters *et al.*, 2016). For these reasons, being phenotypically plastic might be a key trait of successful, i.e. invasive, non-native species. This idea has been the focus of studies for quite some time; its explicit formulation as an invasion hypothesis, however, is more recent (Richards *et al.*, 2006; Davidson *et al.*, 2011; Engel *et al.*, 2011, and

references therein). In the following, we focus on studies that test the (phenotypic) plasticity hypothesis, which posits that invasive species are more phenotypically plastic than non-invasive or native ones,

Previous reviews and synthesis articles on the plasticity hypothesis focused on plants (Daehler, 2003; Richards *et al.*, 2006; Davidson *et al.*, 2011; Palacio-López and Gianoli, 2011). This is the taxonomic group for which the hypothesis was originally formulated (starting with Baker, 1965, reviewed in Richards *et al.* 2006), yet it is equally applicable to other organisms. We here use the hierarchy-of-hypotheses (HoH) approach combined with a systematic literature review across taxonomic groups and a three-level ordinal scoring approach (see Chapters 2 and 6, this volume) to address the following questions: (i) which aspects (i.e. sub-hypotheses) of the plasticity hypothesis have been investigated thus far? (ii) What is the level of support for the overall hypothesis and its sub-hypotheses? (iii) Does the level of support differ among major taxonomic groups, habitats and methodological approaches? (iv) Has the level of support changed over time?

Methods

Systematic literature search

We searched the ISI Web of Science on 17 January 2013, using the following string: 'Phenotypic* AND plastic* AND (alien OR exotic OR introduced OR invasive OR naturali?ed OR nonindigenous OR non-native)'. This search returned 635 hits that we screened by title and abstract. We consulted the full text of those articles that appeared potentially relevant, which finally resulted in 115 relevant empirical studies testing the phenotypic plasticity hypothesis, i.e. that invasive species are more phenotypically plastic than non-invasive or native ones. Purely theoretical tests of the hypothesis were not included, nor reviews or meta-analyses (these were excluded to avoid double-counting of empirical tests).

Hierarchy of hypotheses

Relevant studies were divided into sub-hypotheses and sub-sub-hypotheses according to the type of phenotypic plasticity expressed by the focal non-native species. We hereby used the following categories:

1. Morphology: sub-hypothesis related to morphological phenotypic plasticity, divided into the following categories (i.e. sub-sub-hypotheses): body size, body shape, prey defences, other.
2. Physiology: sub-hypothesis related to physiological phenotypic plasticity, divided into: acclimation, heat shock response, resource-use efficiency, change in chemical composition, organ size.
3. Life history: sub-hypothesis related to phenotypic plasticity in life history, divided into: age of first reproduction, fecundity, development and growth, life span, sex ratio.
4. Behaviour: sub-hypothesis related to behavioural phenotypic plasticity.

Scoring of empirical tests and analysis

We applied the three-level scoring approach as in Jeschke *et al.* (2012a) and Heger and Jeschke (2014), i.e. we categorized the identified relevant empirical tests as either supporting, being undecided or questioning the plasticity hypothesis. As outlined in Chapter 6, this volume, this approach differs from vote counting, which is only based on significance values and has key weaknesses. The approach applied here takes all available evidence into account, particularly effect sizes, to classify studies as supporting, being undecided or questioning. These ordinal scores were used in the further analyses (see below) for which we used the statistical software program SPSS version 21. The dataset

is available freely online at the website www.hi-knowledge.org.

Results and Discussion

Which aspects (i.e. sub-hypotheses) of the plasticity hypotheses have been investigated thus far?

The 115 relevant empirical tests of the plasticity hypothesis that we identified largely fall into three sub-hypotheses: tests that address phenotypic plasticity in: (i) morphology; (ii) physiology; and (iii) life history (Fig. 14.1). The fourth sub-hypothesis – behaviour – was only addressed by few studies, possibly because changes in behaviour are not typically termed 'phenotypic plasticity', even although they fit to the general definition and can be seen as a particularly rapid and reversible form of phenotypic plasticity (Utz *et al.*, 2014).

About one third of the articles had a geographic focus on North America (n = 40), another third on Europe (n = 36) and the final third was spread across the other continents (Asia: n = 13; Australia/Oceania: n = 12; Africa: n = 7; South America: n = 6; Antarctica: n = 1). This geographic bias is not surprising because it has been previously reported by studies looking at the broader field of invasion biology (Pyšek *et al.*, 2008; Bellard and Jeschke, 2016).

What is the level of support for the overall hypothesis and its sub-hypotheses?

Overall, the plasticity hypothesis was supported by the 115 empirical studies in our dataset: 61% (n = 70) were supportive, 23% (n = 27) undecided and 16% (n = 18) were questioning the hypothesis. All three sub-hypotheses with a sufficient amount of available evidence were also supported, and the same was true for most sub-sub-hypotheses (Fig. 14.1).

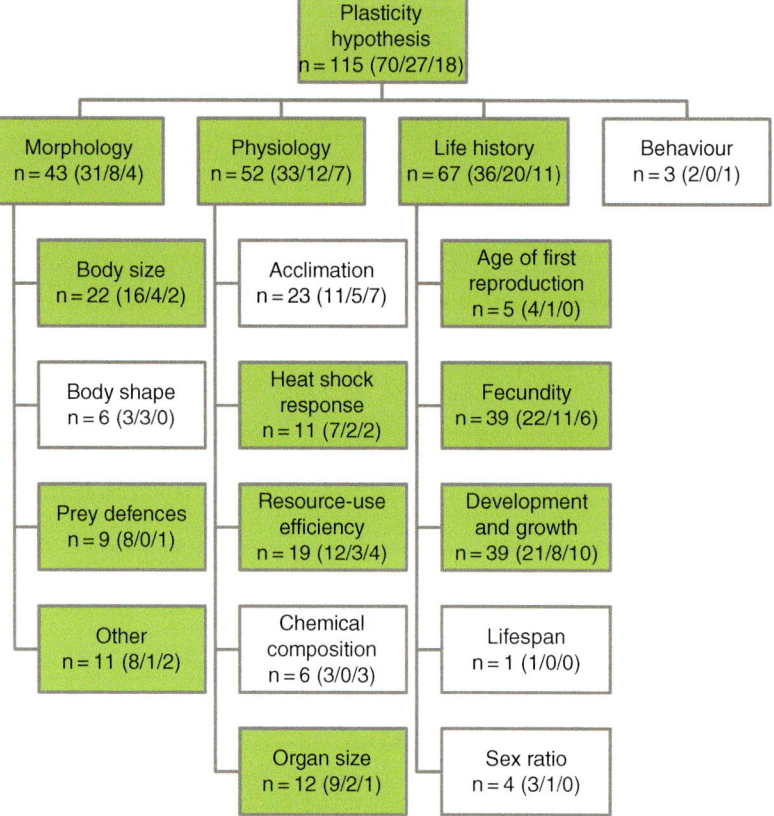

Fig. 14.1. The hierarchy of hypotheses for the phenotypic plasticity hypothesis. The number of empirical studies related to each sub-(sub-)hypothesis add up to more than 115 studies because some studies are related to more than one sub-(sub-)hypothesis. The boxes are colour coded: green indicates that >50% of the empirical studies are supportive and n≥5; red (not existent) would indicate that >50% of the empirical studies are questioning the hypothesis and n≥5; white is used for other cases (i.e. inconclusive data or n<5). Detailed information on the number of studies supporting, being undecided and questioning each (sub-sub-)hypothesis is provided in parentheses, e.g. for the overall hypothesis: 70 studies are supportive, 27 are undecided and 18 are questioning the plasticity hypothesis.

Does the level of support differ among major taxonomic groups, habitats and methodological approaches?

Most studies addressing the plasticity hypothesis were focused on plants, whereas only few studies focused on other organisms (Fig. 14.2a). Although there was a slight tendency for studies on plants to show lower support for the plasticity hypothesis than studies on invertebrates and vertebrates, these differences were not statistically significant (Mann–Whitney U-tests, all two-sided (the same is true for all other significance tests in this chapter): plants vs invertebrates, $p = 0.40$; plants vs vertebrates, $p = 0.32$; invertebrates vs vertebrates, $p = 0.81$). Still, the proportion of studies supporting the plasticity hypothesis in plants was below 60% (it was 58.8%). In comparison, the review studies by Daehler (2003), Davidson *et al.* (2011) and Palacio-López and Gianoli (2011), which were restricted to plants, found rather inconclusive results on the validity of the plasticity hypothesis.

Similarly, most studies on this hypothesis were carried out in terrestrial habitats, followed by freshwater and marine habitats

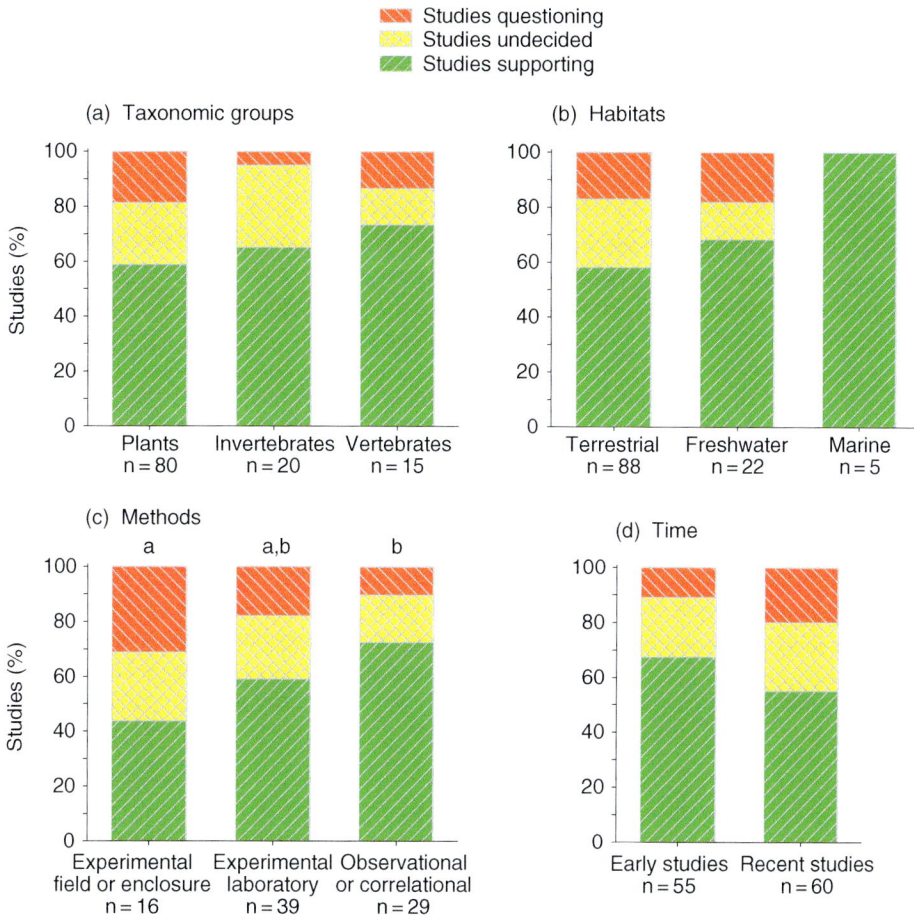

Fig. 14.2. Empirical level of support for the plasticity hypothesis, subdivided for: (a) major taxonomic groups, (b) major habitat types, (c) methodological approaches (here, the number of studies do not add up to 115 because studies using multiple methods were excluded from this comparison) and (d) early vs recent studies. Letters indicate statistically significant differences (U-tests, $p < 0.05$).

(Fig. 14.2b). This is in line with previous synthesis analyses, which showed that most research in invasion biology is in terrestrial habitats and on plants (Pyšek *et al.*, 2008; Jeschke *et al.*, 2012b). Although slight differences do exist among habitats, these are not statistically significant (U-tests: terrestrial vs freshwater, $p = 0.51$; terrestrial vs marine, $p = 0.072$; freshwater vs marine, $p = 0.28$). More than 50% of empirical studies support the plasticity hypothesis in each major type of habitat.

We observed relatively pronounced differences in empirical support among methodological approaches where we compared: (i) experimental field and enclosure (incl. exclosure) studies; (ii) experimental laboratory studies; and (iii) observational or correlational studies (Fig. 14.2c). In the latter category, we did not discriminate between field and enclosure studies on the one hand and laboratory studies on the other hand, because only a few observational studies were carried out in the laboratory (n = 4). Interestingly, experimental field and enclosure studies showed the lowest level of empirical support, followed by experimental laboratory studies and observational studies. The difference between experimental field/enclosure studies and observational studies was statistically significant ($p = 0.045$, U-test; experimental field/enclosure vs experimental laboratory, $p = 0.25$; experimental laboratory vs observational, $p = 0.24$). Thus, those studies that have been designed to mechanistically test for differences showed the lowest level of empirical support, whereas observational studies in which it is harder to filter out effects of single factors (such as phenotypic plasticity) showed higher levels of empirical support. Still, however, slightly more experimental field/enclosure studies were supporting rather than questioning the plasticity hypothesis (44% supporting, 25% being undecided, 31% questioning). All experimental field/enclosure studies were on terrestrial plants and most were after 2010.

Has the level of support changed over time?

A decline in the level of support for six invasion hypotheses was reported by Jeschke et al. (2012a), and decline effects have been previously reported from a few other disciplines, particularly medicine, psychology and evolutionary ecology (Lehrer, 2010; Schooler, 2011). Underlying reasons include publication biases, biases in study organisms or systems and the psychology of researchers (Jeschke et al., 2012a, and references therein).

We divided the empirical studies in our dataset between early studies published until 2010 and recent studies published thereafter, using the cut-off year 2010 to be as close as possible to a 50:50 division between early and recent studies (cf. Jeschke et al., 2012a). Recent studies tended to show lower support for the plasticity hypothesis than earlier studies (Fig. 14.2d); however, this decline was not statistically significant ($p = 0.14$).

Conclusions

The phenotypic plasticity hypothesis is a major invasion hypothesis that is relatively well supported by currently available evidence. Across sub-hypotheses, taxonomic groups, habitats, methodological approaches and time, more studies are supporting than questioning it. Still, less than half of the experimental field and enclosure studies currently available support this hypothesis (many of them are inconclusive). Studies with such a demanding methodological approach are currently rare, hence more are needed to scrutinize the plasticity hypothesis. It would also be good to link behavioural studies (e.g. on flight reactions when predators are approaching or attacks when prey are in reach) to the hypothesis. This has only rarely been done (changes in behaviour can be seen as a particularly rapid and reversible form of phenotypic plasticity; Utz et al., 2014) and would be a promising extension of research focused on the plasticity hypothesis.

References

Baker, H.G. (1965) Characteristics and modes of origin of weeds. In: Baker, H.G. and Stebbins, G.L. (eds) *The Genetics of Colonizing Species*. Academic Press, New York, pp. 147–169.

Bellard, C. and Jeschke, J.M. (2016) A spatial mismatch between invader impacts and research publications. *Conservation Biology* 30, 230–232.

Daehler, C.C. (2003) Performance comparisons of co-occuring native and alien invasive plants: implications for conservation and restoration. *Annual Review of Ecology, Evolution, and Systematics* 34, 183–211.

Davidson, A.M., Jennions, M. and Nicotra, A.B. (2011) Do invasive species show higher phenotypic plasticity than native species and, if so, is it adaptive? A meta-analysis. *Ecology Letters* 14, 419–431.

DeWitt, T.J. and Scheiner, S.M. (eds) (2004) *Phenotypic Plasticity: Functional and Conceptual Approaches*. Oxford University Press, New York.

Engel, K., Tollrian, R. and Jeschke, J.M. (2011) Integrating biological invasions, climate change and phenotypic plasticity. *Communicative & Integrative Biology* 4, 247–250.

Heger, T. and Jeschke, J.M. (2014) The enemy release hypothesis as a hierarchy of hypotheses. *Oikos* 123, 741–750.

Jeschke, J.M., Gómez Aparicio, L., Haider, S., Heger, T., Lortie, C.J., Pyšek, P. and Strayer, D.L. (2012a) Support for major hypotheses in invasion biology is uneven and declining. *Neo-Biota* 14, 1–20.

Jeschke, J.M., Gómez Aparicio, L., Haider, S., Heger, T., Lortie, C.J., Pyšek, P. and Strayer, D.L. (2012b) Taxonomic bias and lack of cross-taxonomic studies in invasion biology. *Frontiers in Ecology and the Environment* 10, 349–350.

Lehrer, J. (2010) The truth wears off. *New Yorker*, Dec 13, 52–57.

Palacio-López, K. and Gianoli, E. (2011) Invasive plants do not display greater phenotypic plasticity than their native or non-invasive counterparts: a meta-analysis. *Oikos* 120, 1393–1401.

Pyšek, P., Richardson, D.M., Pergl, J., Jarošík, V., Sixtová, Z. and Weber, E. (2008) Geographical and taxonomic biases in invasion ecology. *Trends in Ecology & Evolution* 23, 237–244.

Richards, C.L., Bossdorf, O., Muth, N.Z., Gurevitch, J. and Pigliucci, M. (2006) Jack of all trades, master of some? On the role of phenotypic plasticity in plant invasions. *Ecology Letters* 9, 981–993.

Schooler, J. (2011) Unpublished results hide the decline effect. *Nature* 470, 437.

Tollrian, R. and Harvell, C.D. (eds) (1999) *The Ecology and Evolution of Inducible Defenses.* Princeton University Press, Princeton, New Jersey.

Utz, M., Jeschke, J.M., Loeschcke, V. and Gabriel, W. (2014) Phenotypic plasticity with instantaneous but delayed switches. *Journal of Theoretical Biology* 340, 60–72.

Waters, C.N., Zalasiewicz, J., Summerhayes, C., Barnovsky, A.D., Poirier, C., Gałuszka, A., Cearreta, A., Edgeworth, M., Ellis, E.C., Ellis, M. et al. (2016) The Anthropocene is functionally and stratigraphically distinct from the Holocene. *Science* 351, aad2622.

15 Darwin's Naturalization and Limiting Similarity Hypotheses

Jonathan M. Jeschke[1,2,3,4]* and Felix Erhard[4,5]

[1]*Leibniz-Institute of Freshwater Ecology and Inland Fisheries (IGB), Berlin, Germany;* [2]*Freie Universität Berlin, Institute of Biology, Berlin, Germany;* [3]*Berlin-Brandenburg Institute of Advanced Biodiversity Research (BBIB), Berlin, Germany;* [4]*Technical University of Munich, Restoration Ecology, Freising, Germany;* [5]*University of Natural Resources and Life Sciences (BOKU), Vienna, Austria*

Abstract

Darwin's naturalization hypothesis (DN) and the limiting similarity hypothesis (LS) are topically similar: DN posits that the invasion success of non-native species is higher in areas that are poor in closely related species than in areas that are rich in closely related species; LS says that the invasion success of non-native species is high if they strongly differ from native species and low if they are similar to native species. We performed systematic reviews for both DN and LS, and divided these into sub-hypotheses using the hierarchy-of-hypotheses (HoH) approach. For DN, we thereby considered if studies used phylogenies to assess relatedness of native and non-native species or if they did so by using taxonomic groups (e.g. the number of native species in the same *genus* as the non-native species). We found that studies using phylogenies usually support DN, whereas those using taxonomic groups typically question DN. We divided the limiting similarity hypothesis into sub-hypotheses according to how functional similarity was assessed between native and non-native species. This hypothesis was largely empirically supported. Both DN and LS, however, have basically only been addressed in terrestrial habitats, and limiting similarity hypothesis only for plants, thus studies in other habitats and for other taxonomic groups are needed to test the general validity of both hypotheses. Due to their similarity, these hypotheses can be seen as sub-hypotheses of an overarching limiting similarity hypothesis *sensu lato*.

Introduction

Darwin's naturalization hypothesis (DN) is more than a century older than most other invasion hypotheses. It originated from the most prominent biologist, Charles Darwin, who in his famous book *On the Origin of Species* (Darwin, 1859) not only laid the foundation of evolutionary biology but also sowed some seeds for invasion biology (although he was not the only 'early invasion biologist', as outlined by Cadotte, 2006, and Kowarik and Pyšek, 2012). Darwin's naturalization hypothesis posits that the invasion success of non-native species is higher in areas that are poor in closely related species than in areas that are rich in closely related species.

Darwin's naturalization conundrum (Diez *et al.*, 2008) consists, on the one hand,

* Corresponding author. E-mail: jonathan.jeschke@gmx.net

of DN and, on the other hand, of what is now termed the adaptation or pre-adaptation hypothesis, which contradicts DN in that it says: the invasion success of non-native species depends on the adaptation to the conditions in the exotic range before and/or after the introduction; non-native species that are related to native species are more successful in this adaptation (see Chapter 7, this volume; Duncan and Williams, 2002). There are other such contradictory hypotheses (e.g. the biotic resistance and biotic acceptance hypotheses, Chapter 7, this volume), which highlight the need for synthesis as attempted in this book.

The limiting similarity hypothesis (LS) posits that the invasion success of non-native species is high if they strongly differ from native species and low if they are similar to native species. It is based on the same principal rationale as DN, which is easy to see if one uses species relatedness as a proxy for species similarity: doing so, the two hypotheses basically collapse into one hypothesis. Hence, DN can be seen as a sub-hypothesis of the limiting similarity hypothesis *sensu lato*. The limiting similarity hypothesis *sensu stricto* is, however, formulated from a functional trait perspective rather than a phylogenetic perspective. The idea underlying LS was published by MacArthur and Levins (1967), although they did so in general terms and not specified to biological invasions. The limiting similarity hypothesis has slight similarities to the biotic resistance hypothesis (also known as diversity–invasibility hypothesis; Chapter 8, this volume), but should not be confused with it. Publications on how communities can be resistant against invaders from a functional trait perspective have only substantially accumulated after the turn of the millennium (e.g. Symstad, 2000; Emery, 2007; Funk *et al.*, 2008). In summary, we use the following terms to avoid confusion: the limiting similarity hypothesis *sensu lato* is the overarching hypothesis of which DN is a sub-hypothesis; the LS *sensu stricto* – or just LS – is the limiting similarity hypothesis restricted to similarity from a functional trait perspective.

Applying the hierarchy-of-hypotheses (HoH) approach combined with a systematic literature review across taxonomic groups and a three-level ordinal scoring approach for DN and LS (Chapters 2 and 6, this volume; Jeschke *et al.*, 2012a; Heger and Jeschke, 2014), we address the following questions: (i) which aspects (i.e. sub-hypotheses) of DN and LS have been investigated thus far? (ii) What is the level of support for the overall hypotheses and their sub-hypotheses? (iii) Does the level of support differ among major taxonomic groups, habitats and methodological approaches? (iv) Has the level of support changed over time?

Methods

Systematic literature search

For DN, we searched the ISI Web of Science on 26 April 2014, using the following string: '(Darwin* AND (naturali?ation OR hypothesis)) AND (alien OR exotic OR introduc* OR invas* OR naturali?ed OR nonindigenous OR non-indigenous OR nonnative OR non-native)'. For LS, the search was done on 9 May 2014 with the following string: '(Limiting NEAR similarity) AND (alien OR exotic OR introduc* OR invas* OR naturali?ed OR nonindigenous OR non-indigenous OR nonnative OR non-native)'. These searches returned 115 and 157 hits for DN and LS, respectively. We consulted the titles and abstracts of these articles and the full text of those that seemed potentially relevant, i.e. that provided empirical data addressing DN or LS. We also checked references cited in relevant articles and made a forward search in the Web of Science, looking for studies citing relevant articles. We checked if these are relevant for our purposes as well. These systematic searches led to 40 and 33 relevant empirical studies testing DN and LS, respectively. Purely theoretical tests of the hypotheses were not included, nor reviews or meta-analyses (these were excluded to avoid double-counting of empirical tests) but studies cited therein were included if relevant.

Hierarchy of hypotheses

Empirical studies addressing DN were divided into sub-hypotheses according to how relatedness between native and non-native species was measured, discriminating between the following approaches: (i) PNND (phylogenetic nearest neighbour distance) where the phylogenetic distance of the non-native species with the closest native species is measured (Schaefer *et al.*, 2011); (ii) MPD (mean phylogenetic distance), an alternative phylogenetic approach where the mean distance between the non-native species and the investigated native species is calculated (Schaefer *et al.*, 2011; Gallien and Carboni, 2017); (iii) genus membership, a taxonomic approach where the number of native species in the same genus as the non-native species is taken as measurement of relatedness; and (iv) family membership, an alternative taxonomic approach where the number of native species in the same family as the non-native species is counted.

We divided LS into sub-hypotheses according to how functional similarity between native and non-native species was measured, discriminating between: (i) a direct functional traits approach, where functional traits were directly measured by the authors of the study; and (ii) a functional groups approach where functional groups were formed by the authors without actually measuring functional traits (e.g. on the basis of database information about the focal species, which might, however, be inaccurate for the conditions under which the study was performed).

Scoring of empirical tests and analysis

We applied the three-level scoring approach as in Jeschke *et al.* (2012a) and Heger and Jeschke (2014), i.e. we categorized the identified relevant empirical tests as supporting, being undecided or questioning DN and LS. As outlined in Chapter 6, this volume, this approach differs from vote counting, which is only based on significance values and has key weaknesses. The approach applied here takes all available evidence into account, particularly effect sizes, to classify studies as supporting, being undecided or questioning. These ordinal scores were used in the further analyses for which we used the statistical software program SPSS version 21. The dataset is freely available online at the website www.hi-knowledge.org.

Results and Discussion

Which aspects (i.e. sub-hypotheses) of DN and LS have been investigated so far?

We identified 40 relevant empirical studies addressing DN. Some of them used a phylogenetic approach for assessing relatedness between native and non-native species (n = 18); others applied a taxonomic approach (n = 22). The former approach has been used either by means of calculating PNND (phylogenetic nearest neighbour distance, n = 16) or MPD (mean phylogenetic distance, n = 16), while most studies actually used both approaches (n = 14) (Fig. 15.1). The most frequent taxonomic approach in our dataset to measure relatedness is based on the number of native species in the same *genus* as the non-native species. The number of native species in the same *family* as the non-native species has been used less frequently as an estimate for relatedness (Fig. 15.1).

We identified 33 relevant empirical tests of LS. These can be divided into those assessing functional similarity on the basis of actual measurements of functional traits or those assigning native and non-native species to functional groups. The latter approach can be found more frequently in the literature (Fig .15.1), possibly because it is less demanding.

We identified 72 empirical tests of the limiting similarity hypothesis *sensu lato* (DN and LS) because one of the 73 studies addressing DN or LS *sensu stricto* addressed both of these hypotheses (Fig. 15.1). The studies had a strong geographic bias: about half of the studies addressing the limiting similarity hypothesis *sensu lato* focused on North America, 15% each on Europe and

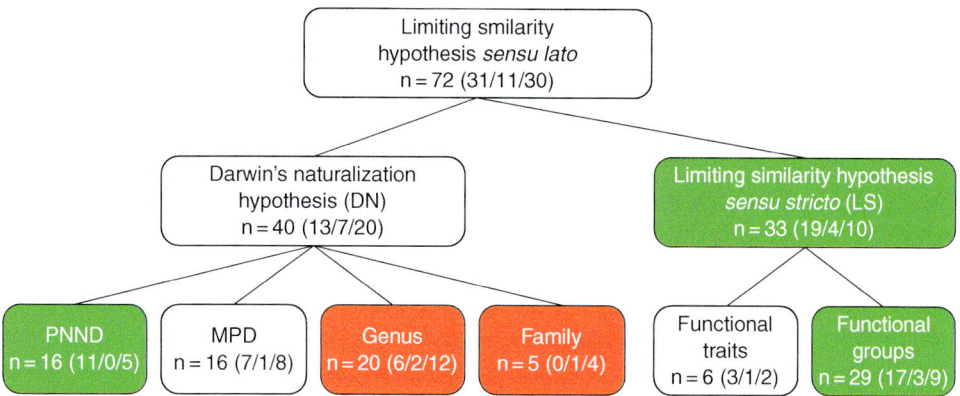

Fig. 15.1.   The hierarchy of hypotheses for the limiting similarity hypothesis *sensu lato*, divided into DN and LS. The number of empirical studies related to sub-hypotheses of DN and LS add up to more than 40 and 33 studies, respectively, because some studies are related to more than one sub-hypothesis. Similarly, one study addressed both DN and LS, which is why the number of studies addressing the limiting similarity hypothesis *sensu lato* is 72 rather than 73. The boxes are colour coded: green indicates that >50% of the empirical studies are supportive and $n \geq 5$; red indicates that >50% of the empirical studies are questioning the hypothesis and $n \geq 5$; white is used for other cases (i.e. inconclusive data or $n < 5$). Detailed information on the number of studies supporting, being undecided and questioning each (sub-sub-) hypothesis is provided in parentheses, e.g. for DN: 13 studies are supportive, 7 are undecided and 20 are questioning DN.

Australia/Oceania, and the remaining studies on other or multiple continents. Such a geographic bias is not unusual in invasion biology (Pyšek *et al.*, 2008).

What is the level of support for the overall hypotheses and their sub-hypotheses?

Darwin's naturalization hypothesis received significantly less empirical support than the limiting similarity hypothesis *sensu stricto* ($p = 0.039$, n = 71, the one study addressing both hypotheses was excluded here; Mann–Whitney U-test, two-sided (the same is true for all other significance tests in this chapter)). Consequently, the combined limiting similarity hypothesis *sensu lato* received mixed support (Fig. 15.1).

Although there was no strong difference between the two sub-hypotheses of LS, strong differences were observed between the sub-hypotheses of DN (Fig. 15.1): the majority of studies applying PNND reported supporting evidence for DN, whereas those using MPD reported mixed evidence. This was particularly interesting because most

publications used both of these approaches, hence these differences between PNND and MPD are not due to differences in study design or environmental conditions but are actual differences between the two approaches. The majority of studies applying a taxonomic approach (genus or family level) questioned DN. These differences were (marginally) statistically significant (Mann–Whitney U-tests: phylogenetic studies (n = 18) vs taxonomic studies (n = 22), $p = 0.065$; PNND (n = 16) vs genus (n = 20), $p = 0.041$). Our results indicate that DN should be applied from a phylogenetic rather than taxonomic perspective and that the key phylogenetic aspect is the distance of the focal non-native species to the *closest* native species (i.e. PNND) rather than the mean phylogenetic distance (MPD).

Does the level of support differ among major taxonomic groups, habitats and methodological approaches?

Both DN and LS have basically only been addressed in terrestrial habitats, and mostly for terrestrial plants (Figs 15.2 and 15.3). There are also some studies for vertebrates on DN but not on LS, and basically no studies for other organisms. This bias regarding taxonomy and habitat is stronger than for other major invasion hypotheses (Chapters 8–17, this volume; Jeschke *et al.*, 2012a,b) and in the general field of invasion biology (Pyšek *et al.*, 2008). There was no apparent difference in the level of support of DN between studies on plants and vertebrates (Fig. 15.2a).

Studies can be divided and compared in several ways when it comes to differences in the methodological approach. In fact, we used one methodological aspect to structure our HoH for both DN and LS, thus for the overarching limiting similarity hypothesis

sensu lato, and found significant differences (Fig. 15.1); hence it is key how to measure species similarity and relatedness. Another possibility is to compare experimental vs observational/correlational studies. More observational than experimental studies have been done for DN, whereas more experimental than observational studies have been done on LS (Figs 15.2 and 15.3). There were no statistically significant differences between experimental and observational studies for either of the two hypotheses (U-tests; DN: $p = 0.56$; LS: $p = 0.98$). Thus for these hypotheses, experiments seem to have covered the relevant processes occurring in the field and, vice versa, observational studies seem to be as reliable in their results as experiments.

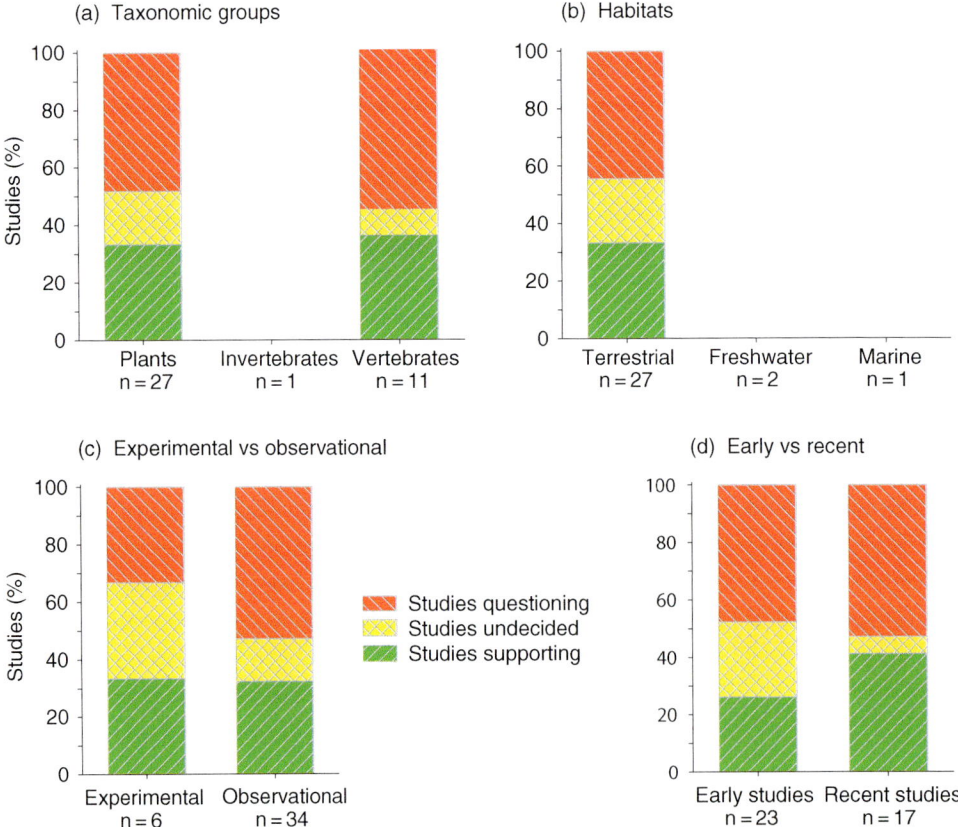

Fig. 15.2. Empirical level of support for Darwin's naturalization hypothesis (DN), subdivided for (a) major taxonomic groups, (b) major habitat types (here, the number of studies do not add up to 40 because studies in multiple habitats were excluded from this comparison), (c) experimental vs observational methodological approaches and (d) early vs recent studies. There were no statistically significant differences.

Has the level of support changed over time?

A decline in the level of support for six invasion hypotheses was reported by Jeschke et al. (2012a), and decline effects have been previously reported from a few other disciplines (Lehrer, 2010; Schooler, 2011). Possible underlying reasons are publication biases, biases in study organisms or systems and the psychology of researchers (Jeschke et al., 2012a, and references therein).

We divided the empirical studies in our dataset between early studies and recent studies, using a cut-off year for DN and LS to be as close as possible to a 50:50 division between early and recent studies for these hypotheses (cf. Jeschke et al., 2012a). The cut-off year was 2011 for DN (i.e. early studies were published until 2011, recent studies thereafter) and 2008 for LS. There were no relevant differences in empirical support between early and recent studies in DN or LS (Figs 15.2d and 15.3d), hence a decline effect was not observed.

Conclusions

The limiting similarity hypothesis is largely empirically supported by currently available studies, and Darwin's naturalization hypothesis has been supported by studies using a phylogenetic approach. In contrast, studies applying a taxonomic approach have usually questioned DN, and the difference between phylogenetic and taxonomic approaches

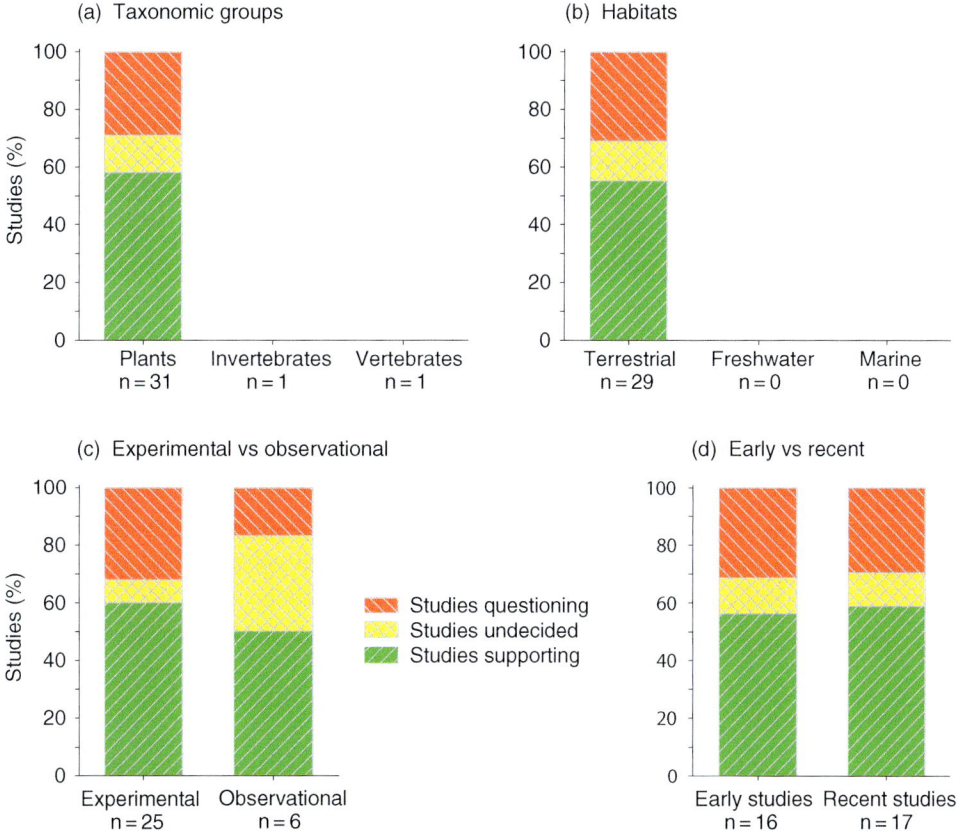

Fig. 15.3. As for Fig. 15.2, but for the limiting similarity hypothesis *sensu stricto* (LS). There were no statistically significant differences here either.

seems clear enough to suggest a specification of DN that it refers to relatedness from a phylogenetic perspective rather than a taxonomic perspective. It remains to be seen, however, whether DN from this perspective is widely applicable because most currently available tests are restricted to terrestrial habitats and to plants, hence studies in other habitats and for other groups of organisms are highly needed. Available evidence for LS also mainly stems from terrestrial plants. We currently do not know if it also holds for other habitats and other groups of organisms.

References

Cadotte, M.W. (2006) Darwin to Elton: early ecology and the problem of invasive species. In: Cadotte, M.W., McMahon, S.M. and Fukami, T. (eds) *Conceptual Ecology and Invasion Biology: Reciprocal Approaches to Nature.* Springer, Dordrecht, Netherlands, pp. 15–33.

Darwin, C. (1859) *On the Origin of Species by Means of Natural Selection.* Murray, London.

Diez, J.M., Sullivan, J.J., Hulme, P.E., Edwards, G. and Duncan, R.P. (2008) Darwin's naturalization conundrum: dissecting taxonomic patterns of species invasions. *Ecology Letters* 11, 674–681.

Duncan, R.P. and Williams, P.A. (2002) Darwin's naturalization hypothesis challenged. *Nature* 417, 608–609.

Emery, S.M. (2007) Limiting similarity between invaders and dominant species in herbaceous plant communities? *Journal of Ecology* 95, 1027–1035.

Funk J.L., Cleland E.E., Suding K.N. and Zavaleta E.S. (2008) Restoration through reassembly: plant traits and invasion resistance. *Trends in Ecology & Evolution* 23, 695–703.

Gallien, L. and Carboni, M. (2017) The community ecology of invasive species: where are we and what's next? *Ecography* 40, 335–352.

Heger, T. and Jeschke, J.M. (2014) The enemy release hypothesis as a hierarchy of hypotheses. *Oikos* 123, 741–750.

Jeschke, J.M., Gómez Aparicio, L., Haider, S., Heger, T., Lortie, C.J., Pyšek, P. and Strayer, D.L. (2012a) Support for major hypotheses in invasion biology is uneven and declining. *NeoBiota* 14, 1–20.

Jeschke, J.M., Gómez Aparicio, L., Haider, S., Heger, T., Lortie, C.J., Pyšek, P. and Strayer, D.L. (2012b) Taxonomic bias and lack of cross-taxonomic studies in invasion biology. *Frontiers in Ecology and the Environment* 10, 349–350.

Kowarik, I. and Pyšek, P. (2012) The first steps towards unifying concepts in invasion ecology were made one hundred years ago: revisiting the work of the Swiss botanist Albert Thellung. *Diversity and Distributions* 18, 1243–1252.

Lehrer, J. (2010) The truth wears off. *New Yorker* Dec 13, 52–57.

MacArthur, R. and Levins, R. (1967) The limiting similarity, convergence, and divergence of coexisting species. *American Naturalist* 377–385.

Pyšek, P., Richardson, D.M., Pergl, J., Jarošík, V., Sixtová, Z. and Weber, E. (2008) Geographical and taxonomic biases in invasion ecology. *Trends in Ecology & Evolution* 23, 237–244.

Schaefer, H., Hardy, O.J., Silva, L., Barraclough, T.G. and Savolainen, V. (2011) Testing Darwin's naturalization hypothesis in the Azores. *Ecology Letters* 14, 389–396.

Schooler, J. (2011) Unpublished results hide the decline effect. *Nature* 470, 437–437.

Symstad, A.J. (2000) A test of the effects of functional group richness and composition on grassland invasibility. *Ecology* 81, 99–109.

16 Propagule Pressure Hypothesis

Jonathan M. Jeschke[1,2,3,4]* and Julian Starzer[4]

[1]Leibniz-Institute of Freshwater Ecology and Inland Fisheries (IGB), Berlin, Germany; [2]Freie Universität Berlin, Institute of Biology, Berlin, Germany; [3]Berlin-Brandenburg Institute of Advanced Biodiversity Research (BBIB), Berlin, Germany; [4]Technical University of Munich, Restoration Ecology, Freising, Germany

Abstract

Propagule pressure is a composite measure of introduction effort consisting of: (i) the number of individuals introduced per introduction event (propagule size); and (ii) the frequency of introduction events (propagule frequency or number). The propagule pressure hypothesis posits that a high propagule pressure is a cause of invasion success; in other words, non-native species with a high propagule pressure have a higher invasion success than non-native species with a low propagule pressure. On the basis of a systematic review, we identified 92 relevant empirical studies testing the propagule pressure hypothesis. These studies have addressed different aspects – that is, sub- and sub-sub-hypotheses – of the overall hypothesis. Independently of the specific aspects considered by each study, the propagule pressure hypothesis is largely supported by currently available evidence. About 80% of the 92 studies reported supporting evidence. Similarly, the propagule pressure hypothesis is empirically supported across major taxonomic groups, habitats and methodological approaches. This hypothesis is among the most influential ones in the field and represents the recognition that in order to understand biological invasions, one must consider humans and their actions as key underlying drivers.

Introduction

Propagule pressure has become an important term in invasion biology after the turn of the millennium. Richardson (2004), p. 317, wrote: 'understanding propagule pressure is the new frontier in invasion ecology'. It is a measure of introduction effort for a given non-native species and is often divided into two key components: (i) propagule size, which is the number of individuals of the non-native species introduced per introduction event; and (ii) propagule frequency, which is the frequency of introduction events. Alternatively to the latter, propagule number is often used, which is the number of introduction events in a given time period (Lockwood *et al.*, 2005, 2009, 2013; Blackburn *et al.*, 2009; Simberloff, 2009; Jeschke, 2014; Wittmann *et al.*, 2014).

The propagule pressure hypothesis posits that a high propagule pressure is a cause of invasion success (Lockwood *et al.*, 2005). In other words, it says that non-native species with a high propagule pressure have a higher invasion success than non-native species with a low propagule pressure. Or in

* Corresponding author. E-mail: jonathan.jeschke@gmx.net

simple terms: the more you introduce, the more you get (cf. Lockwood *et al.*, 2009). This seems obvious but there are possible contrary effects, e.g. an increasing proportion of maladapted individuals with increasing propagule size countering adaptation to local conditions.

The propagule pressure hypothesis can arguably be seen as the most important hypothesis of the field right now, as indicated by the above quote from Richardson (2004) and a recent survey by ourselves where the enemy release and propagule pressure hypotheses were indicated by the relative majority of the >350 experts taking the survey to be the hypotheses they know best (Enders *et al.*, 2018). In contrast, the systematic review by Lowry *et al.* (2013) found that a few other hypotheses were the focus of more publications. Yet, that review covered literature published from the past until 2011, and the propagule pressure hypothesis has only become popular in the 21st century.

The propagule pressure hypothesis captures two important insights that have only become clear to invasion biologists after a significant body of research had become available. The first one is that biological invasions can only be understood if the role of human action is explicitly considered. It is humans who introduce non-native species, either intentionally or unintentionally, thus propagule pressure is a direct result of human action. The second important insight is that event-level characteristics play an important role in biological invasions, i.e. factors associated with, and often unique to, introduction events (Duncan *et al.*, 2003; Blackburn *et al.*, 2009). In the early days of invasion biology, most research was devoted to either specific traits that potentially increase invasiveness of non-native species (species-level characteristics) or traits of environments potentially reducing their invasibility (location-level characteristics) (e.g. Drake *et al.*, 1989); event-level characteristics and human action were not so much in the focus. The rise in importance of the propagule pressure hypothesis indicates that this has changed.

Although this hypothesis is widely known and acknowledged in the field, some

researchers have questioned its validity (Moulton *et al.*, 2011, 2013; rebuttals by Blackburn *et al.*, 2011, 2013). It is possible, of course, that species- and location-level characteristics are much more important than propagule pressure. This chapter aims to shed light on this issue. More specifically, we use the hierarchy-of-hypotheses (HoH) approach combined with a systematic literature review across taxonomic groups and a three-level ordinal scoring approach to address the following questions: (i) which aspects (sub-hypotheses) of the propagule pressure hypothesis have been investigated so far? (ii) What is the level of support for the overall hypothesis and its sub-hypotheses? (iii) Does the level of support differ among major taxonomic groups, habitats and methodological approaches? (iv) Has the level of support changed over time?

Methods

Systematic literature search

We searched the ISI Web of Science on 27 April 2014 with the following term: 'propagule* AND (alien OR exotic OR introduc* OR invas* OR naturali?ed OR nonindigenous OR non-indigenous OR nonnative OR non-native)'. This search returned 1437 hits that we screened for relevance, first identifying potentially relevant articles based on titles and abstracts and then using the full text of these in order to identify relevant empirical studies addressing the propagule pressure hypothesis. We found 81 relevant papers in this way. We also checked papers cited in relevant papers (backward search) and made a forward search in the Web of Science for papers citing those articles that we classified as relevant. These searches yielded 11 additional relevant papers; thus we identified a total of 92 relevant empirical studies addressing the propagule pressure hypothesis. Purely theoretical tests of the hypothesis were not included, nor reviews or meta-analyses (these were excluded to avoid double-counting of empirical tests).

Hierarchy of hypotheses

Applying the HoH approach, we divided the relevant studies into sub- and sub-sub-hypotheses according to how propagule pressure and invasion success were measured. More specifically, we considered the following categories:

- Sub-hypotheses were formed on the basis of how propagule pressure was estimated: (i) total number released, which is the total number of individuals of the non-native species released to an area over time; (ii) propagule size, which is the number of individuals of the non-native species introduced per introduction event; (iii) propagule frequency (the frequency of introduction events) or propagule number (the number of introduction events in a given time period); (iv) distance from source as a proxy for propagule pressure, for instance from cities, gardens, plantations, roads or other invasion foci; and (v) various other proxies, e.g. the amount of trade and sales or human population density.
- Within each sub-hypothesis, sub-sub-hypotheses were formed on the basis of how invasion success was measured: (i) abundance, biomass or cover at the level of populations; (ii) establishment success vs failure of the non-native species, also at the population level; (iii) survival, growth or reproduction at the level of individuals; and (iv) other measures, e.g. spread and colonization success.

Scoring of empirical tests and analysis

We applied the three-level scoring approach as in Jeschke *et al.* (2012a) and Heger and Jeschke (2014; Chapter 2, this volume), i.e. we categorized the identified relevant empirical tests as either supporting, being undecided or questioning the propagule pressure hypothesis. As outlined in Chapter 6, this volume, this scoring approach is different from vote counting, which is only based on significance values and has key weaknesses. The scoring approach applied here takes all available evidence into account, particularly effect sizes, to classify studies as supporting (i.e. results were in the direction predicted by the hypothesis), questioning (i.e. results were in the opposite direction or no effect was shown) or being undecided (i.e. results were partly supporting and partly questioning, e.g. for different sites covered by a study). These ordinal scores were used in the further analyses for which we used the statistical software program SPSS version 21. The dataset is freely available online at www.hi-knowledge.org.

Results and Discussion

Which aspects (sub-hypotheses) of the propagule pressure hypothesis have been investigated thus far?

More than 40% of the studies on the propagule pressure hypothesis focused on propagule size as measure of propagule pressure, i.e. these studies looked at the number of individuals of the non-native species introduced per introduction event (Fig. 16.1). About 15% of the studies each focused on the total number of individuals released or on propagule frequency (i.e. the frequency of introduction events) as measures of propagule pressure. Distance from source was the most frequently used proxy for propagule pressure: it was applied by about 15% of the studies in our dataset. About another 15% of the studies used other proxies (Fig. 16.1).

Of the measures of invasion success, the most frequent one in our dataset was establishment success: across measures of propagule pressure, 36 studies focused on establishment success. Its combination with the most frequently used measure of propagule pressure (i.e. propagule size, see previous paragraph) led to the most frequent sub-sub-hypothesis, namely that non-native species with a high propagule size have a higher establishment success than non-native species with a low propagule size.

About 20% of the studies in our dataset tested this sub-sub-hypothesis. Some other sub-sub-hypotheses were also tested by a respectable number of studies (Fig. 16.1).

There was a geographic bias among the studies in our dataset: more than 30% of the studies focused on North America (n = 36), about 20% each on Australia/Oceania (n = 25) and Europe (n = 22), and the remaining studies were in other continents (Africa n = 15, South America n = 7, Asia n = 6 and Antarctica n = 1). Similar geographic biases have been reported in previous studies (Pyšek et al., 2008; Bellard and Jeschke, 2016) and other chapters in this book.

What is the level of support for the overall hypotheses and their sub-hypotheses?

The propagule pressure hypothesis is supported by most of the studies in our dataset. About 80% of the studies reported supporting evidence (Fig. 16.1). This empirical support basically covers all sub- and sub-sub-hypotheses. Comparing these results to those in other chapters of this book, the propagule pressure hypothesis seems to be one of few invasion hypotheses where currently available evidence is clearly supportive.

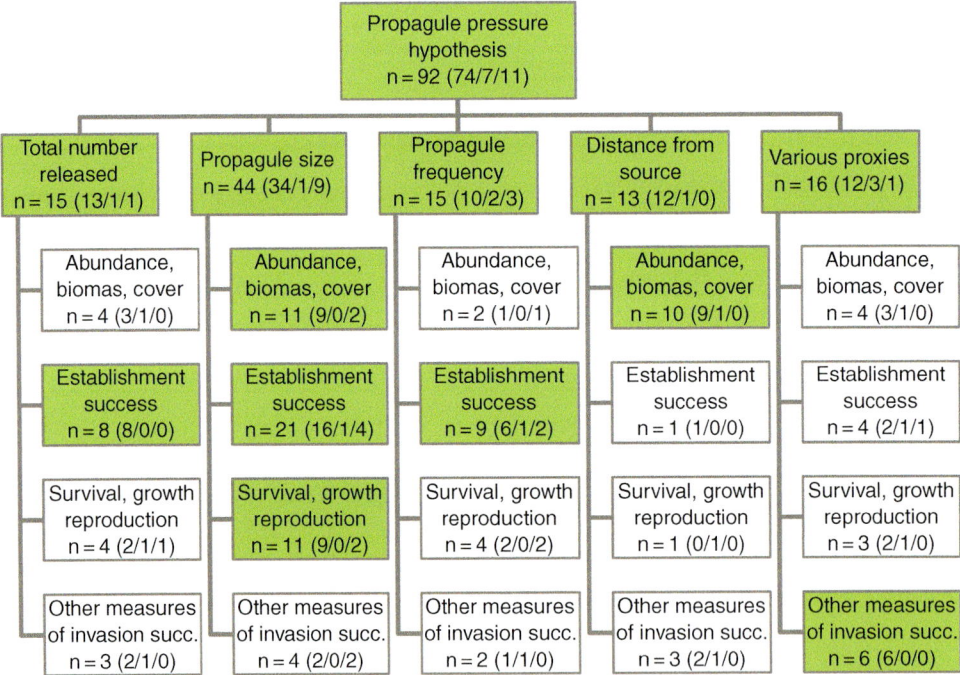

Fig. 16.1. The hierarchy of hypotheses for the propagule pressure hypothesis. The number of empirical studies related to each sub-(sub-)hypothesis adds up to more than 92 studies because some studies are related to more than one sub-(sub-)hypothesis. The boxes are colour coded: green indicates that >50% of the empirical studies are supportive and n≥5; red (not existent) would indicate that >50% of the empirical studies are questioning the hypothesis and n≥5; and white is used for other cases (i.e. inconclusive data or n<5). Detailed information on the number of studies supporting, being undecided and questioning each (sub-sub-)hypothesis is provided in parentheses, e.g. for the overall hypothesis: 74 studies are supportive, 7 are undecided and 11 are questioning the propagule pressure hypothesis.

Does the level of support differ among major taxonomic groups, habitats and methodological approaches?

The relative majority of studies (45%) on the propagule pressure hypothesis focused on non-native plants but high numbers of studies also looked at invertebrates (22%) and vertebrates (33%; Fig. 16.2). Thus, there is a weaker taxonomic bias among studies on this hypothesis than in many other invasion hypotheses (Jeschke *et al.*, 2012b, and chapters in this book). There were no statistically significant differences in the level of support among taxonomic groups (Mann–Whitney U-tests, all two-sided (the same is true for all other significance tests in this chapter):

plants vs invertebrates, $p = 0.12$; plants vs vertebrates, $p = 0.25$; invertebrates vs vertebrates, $p = 0.63$).

We observed a strong bias towards terrestrial habitats where 78% of the studies were performed (Fig. 16.2); 18% of the studies were done in freshwaters and only 5% in marine habitats. There was no statistically significant difference in empirical support between terrestrial and freshwater studies ($p = 0.47$; due to the low number of marine studies, we performed no significance tests for these studies).

About a third of the studies in the dataset followed an experimental approach, whereas two-thirds of the studies were observational (Fig. 16.2). There was basically

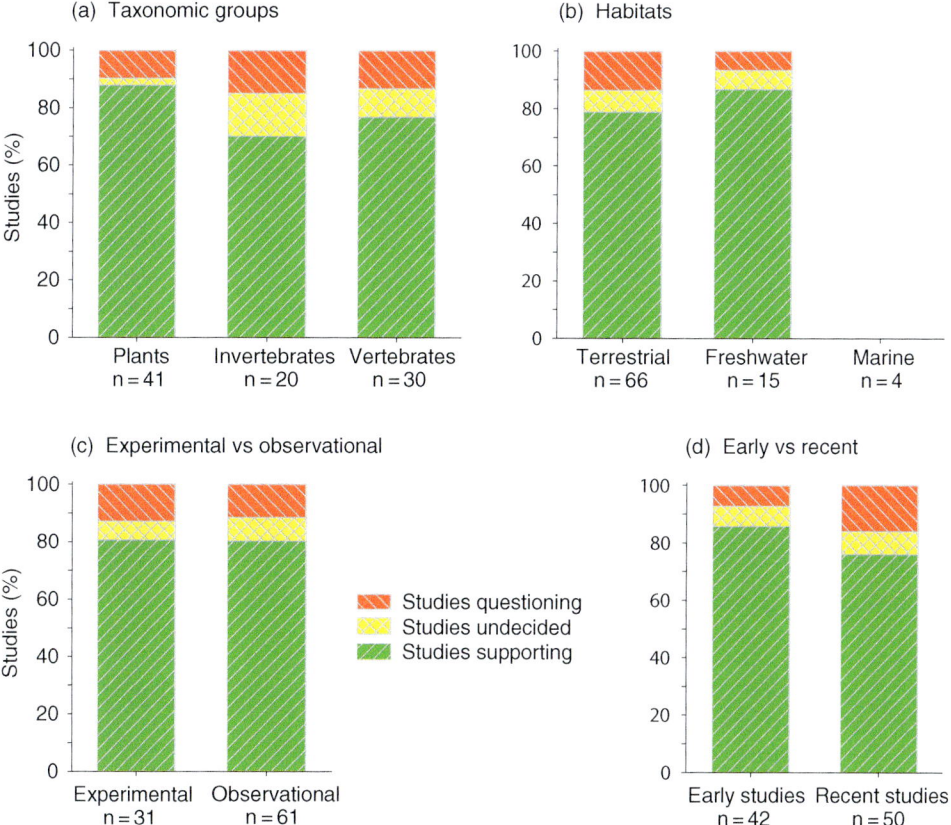

Fig. 16.2. Empirical level of support for the propagule pressure hypothesis, subdivided for (a) major taxonomic groups, (b) major habitat types (studies in multiple habitats were excluded from this comparison), (c) experimental vs observational methodological approaches and (d) early (until 2008) vs recent (since 2009) studies. There were no statistically significant differences.

no difference in the level of empirical support for the propagule pressure hypothesis between experimental and observational approaches.

Has the level of support changed over time?

We divided the empirical studies in our dataset between early and recent studies, using the cut-off year 2008 in order to be as close as possible to a 50:50 division between early and recent studies (cf. Jeschke *et al.*, 2012a). There was a slight decrease in empirical support over time between studies published until 2008 (early studies) and those published since 2009 (recent studies); however, this slight decrease was not statistically significant ($p = 0.22$; Fig. 16.2).

Outlook

As outlined in this chapter, the propagule pressure hypothesis and its sub-(sub-) hypotheses are well supported across taxonomic groups, habitats and methodological approaches. This hypothesis is a modern invasion hypothesis that has only become prominent in the 21st century. Its comparably high level of empirical support indicates that explanation and prediction of biological invasions can be enhanced if the influence of human actions is explicitly considered. Modern invasion biology is a truly interdisciplinary field that stretches far into social sciences. For this reason, Richardson and Ricciardi (2013) suggested to use the term invasion *science* rather than invasion biology or ecology, highlighting its interdisciplinary nature. We suggest to further strengthen interdisciplinary and transdisciplinary work in this field. Such work is sometimes challenged by differing perspectives and terminologies among research disciplines, and between researchers and other stakeholders (Heger *et al.*, 2013; Jeschke *et al.*, 2014; Courchamp *et al.*, 2017). It is high time to efficiently tackle these challenges.

References

Bellard, C. and Jeschke, J.M. (2016) A spatial mismatch between invader impacts and research publications. *Conservation Biology* 30, 230–232.

Blackburn, T.M., Lockwood, J.L. and Cassey, P. (2009) Avian invasions: the ecology and evolution of exotic bird species. Oxford University Press, Oxford, UK.

Blackburn, T.M., Prowse, T.A.A., Lockwood, J.L. and Cassey, P. (2011) Passerine introductions to New Zealand support a positive effect of propagule pressure on establishment success. *Biodiversity and Conservation* 20, 2189–2199.

Blackburn, T.M., Prowse, T.A.A., Lockwood, J.L. and Cassey, P. (2013) Propagule pressure as a driver of establishment success in deliberately introduced exotic species: fact or artefact? *Biological Invasions* 15, 1459–1469.

Courchamp, F., Fournier, A., Bellard, C., Bertelsmeier, C., Bonnaud, E., Jeschke, J.M. and Russell, J.C. (2017) Invasion biology: specific problems and possible solutions. *Trends in Ecology & Evolution* 32, 13–22.

Drake, J.A., Mooney, H.A., di Castri, F., Groves, R.H., Kruger, F.J., Rejmánek, M. and Williamson, M. (eds) (1989) *Biological Invasions: a Global Perspective* – SCOPE 37. Wiley, Chichester, UK.

Duncan, R.P., Blackburn, T.M. and Sol, D. (2003) The ecology of bird introductions. *Annual Review of Ecology, Evolution, and Systematics* 34, 71–98.

Enders, M., Hütt, M.-T. and Jeschke, J.M. (2018) Drawing a map of invasion biology based on a network of hypotheses. *Ecosphere*. DOI: 10.1002/ecs2.2146.

Heger, T. and Jeschke, J.M. (2014) The enemy release hypothesis as a hierarchy of hypotheses. *Oikos* 123, 741–750.

Heger, T., Pahl, A.T., Botta-Dukat, Z., Gherardi, F., Hoppe, C., Hoste, I., Jax, K., Lindström, L., Boets, P., Haider, S. *et al.* (2013) Conceptual frameworks and methods for advancing invasion ecology. *Ambio* 42, 527–540.

Jeschke, J.M. (2014) General hypotheses in invasion ecology. *Diversity and Distributions* 20, 1229–1234.

Jeschke, J.M., Gómez Aparicio, L., Haider, S., Heger, T., Lortie, C.J., Pyšek, P. and Strayer, D.L. (2012a) Support for major hypotheses in invasion biology is uneven and declining. *NeoBiota* 14, 1–20.

Jeschke, J.M., Gómez Aparicio, L., Haider, S., Heger, T., Lortie, C.J., Pyšek, P. and Strayer,

D.L. (2012b) Taxonomic bias and lack of cross-taxonomic studies in invasion biology. *Frontiers in Ecology and the Environment* 10, 349–350.

Jeschke, J.M., Bacher, S., Blackburn, T.M., Dick, J.T.A., Essl, F., Evans, T., Gaertner, M., Hulme, P.E., Kühn, I., Mrugała, A. *et al.* (2014) Defining the impact of non-native species. *Conservation Biology*, 28, 1188–1194.

Lockwood, J.L., Cassey, P. and Blackburn, T. (2005) The role of propagule pressure in explaining species invasions. *Trends in Ecology & Evolution* 20, 223–228.

Lockwood, J.L., Cassey, P. and Blackburn, T. (2009) The more you introduce the more you get: the role of colonization pressure and propagule pressure in invasion ecology. *Diversity and Distributions* 15, 904–910.

Lockwood, J.L., Hoopes, M.F. and Marchetti, M.P. (2013) *Invasion Ecology*, 2nd edn. Wiley-Blackwell, Chichester, UK.

Lowry, E., Rollinson, E.J., Laybourn, A.J., Scott, T.E., Aiello-Lammens, M.E., Gray, S.M., Mickley, J. and Gurevitch, J. (2013) Biological invasions: a field synopsis, systematic review, and database of the literature. *Ecology and Evolution* 3, 182–196.

Moulton, M.P., Cropper, W.P., Jr and Avery, M.L. (2011) A reassessment of the role of propagule pressure in influencing fates of passerine introductions to New Zealand. *Biodiversity and Conservation* 20, 607–623.

Moulton, M.P., Cropper, W.P., Jr and Avery, M.L. (2013) Is propagule size the critical factor in predicting introduction outcomes in passeriform birds? *Biological Invasions* 15, 1449–1458.

Pyšek, P., Richardson, D.M., Pergl, J., Jarošík, V., Sixtová, Z. and Weber, E. (2008) Geographical and taxonomic biases in invasion ecology. *Trends in Ecology & Evolution* 23, 237–244.

Richardson, D.M. (2004) Plant invasion ecology – dispatches from the front line. *Diversity and Distributions* 10, 315–319.

Richardson, D.M. and Ricciardi, A. (2013) Misleading criticisms of invasion science: a field guide. *Diversity and Distributions* 19, 1461–1467.

Simberloff, D. (2009) The role of propagule pressure in biological invasions. *Annual Review of Ecology, Evolution, and Systematics* 40, 81–102.

Wittmann, M.J., Metzler, D., Gabriel, W. and Jeschke, J.M. (2014) Decomposing propagule pressure: the effects of propagule size and propagule frequency on invasion success. *Oikos* 123, 441–450.

Part III

Synthesis and Outlook

17 Synthesis

Jonathan M. Jeschke[1,2,3]* and Tina Heger[4,5,3]

[1]Leibniz-Institute of Freshwater Ecology and Inland Fisheries (IGB), Berlin, Germany; [2]Freie Universität Berlin, Institute of Biology, Berlin, Germany; [3]Berlin-Brandenburg Institute of Advanced Biodiversity Research (BBIB), Berlin, Germany; [4]University of Potsdam, Biodiversity Research/Systematic Botany, Potsdam, Germany; [5]Technical University of Munich, Restoration Ecology, Freising, Germany

Abstract

About 1100 studies focusing on 12 major invasion hypotheses have been analysed in Chapters 8–16 of this book. A network of these 12 hypotheses, in which topically similar hypotheses are connected, was presented in Chapter 7. We here combine and synthesize these previous chapters, colour coding the hypothesis network depending on the level of empirical support of each hypothesis. Overall, six of the 12 hypotheses were supported by the majority of available empirical studies, three hypotheses were questioned by the majority of studies, and empirical studies were undecided for the three remaining hypotheses. The three questioned hypotheses were: evolution of increased competitive ability (EICA), biotic resistance and the tens rule. On the basis of these findings, we propose an alternative hypothesis network in which the biotic resistance hypothesis and the tens rule are replaced by revised hypotheses that are better empirically supported, and the EICA hypothesis is abandoned because the better empirically supported shifting defence hypothesis already is a refinement of this hypothesis. The revised hypothesis network therefore consists of 11 major hypotheses. Most studies analysed in this book focused on terrestrial plants in affluent countries, whereas other taxonomic groups, habitats and other countries are underrepresented in the invasion literature. Observational studies currently dominate the field. We further found that the level of empirical support has declined over time for some but not all focal hypotheses. The hypothesis network featured here is provided online, where it is also connected to the empirical data analysed in this book. This website is envisioned as the initiation of an advanced online tool that grows beyond invasion biology and should cover different scientific disciplines such as community ecology, biodiversity science and evolutionary biology. It is meant to visualize the major concepts, ideas and hypotheses in these disciplines including their links and connections, thus featuring a large structured network that is connected to the data generated in these disciplines.

Science and Society Need New Tools for Research Synthesis

Even when informed and well-intentioned scientists try to think broadly about research options, their discussions suffer from the absence of a synthetic vision.

---

* Corresponding author. E-mail: jonathan.jeschke@gmx.net

Instead of pitting one partial perspective against another, it would be preferable to create a space in which the entire range of our inquiries could be soberly appraised. We would do well to have an institution for the construction and constant revision of an atlas of scientific significance.

(p. 127 in Kitcher, 2011)

These lines are from Kitcher's highly recommended book *Science in a Democratic Society* in which he points out the dwindling importance of science in today's societies. This book was written and published before the more recent developments in, for example, the UK (the Brexit vote in 2016) and the USA (the 2016 Presidential elections) that have led to discussions about post-truth societies. The term 'post-truth' was the Oxford Dictionaries Word of the Year 2016, an adjective defined as 'relating to or denoting circumstances in which objective facts are less influential in shaping public opinion than appeals to emotion and personal belief' (https://en.oxforddictionaries.com/word-of-the-year/word-of-the-year-2016). Since the Trump administration has been in office, euphemisms such as 'alternative facts' have been used more frequently.

What can science do in such times? It can and has to do a lot. As pointed out by Kitcher (2011), one thing science can do is to develop innovative solutions. Expanding on his arguments, we suggest that new synthesis tools such as an 'atlas of scientific significance' can: (i) improve the transparency of scientific claims, for peer scientists as well as non-scientists; (ii) reduce the delay between the times when scientific findings are made and when they are being accessible to others; and (iii) correct for research biases owing to financial or socio-political interests that researchers might have (e.g. Lokatis and Jeschke, 2018). Such new synthesis tools should complement open science (e.g. https://osf.io) and other initiatives associated with transparency and reproducibility to help science strengthen its role in today's societies.

But which tools do we exactly mean and how should they differ from classical synthesis tools? Research has traditionally been synthesized in textbooks or review papers, more recently by way of systematic reviews and meta-analyses. What you have in your hands right now is a book as well (or maybe an electronic device with which you are reading this book) and we do not at all mean to stop writing books or papers; otherwise this book would not exist. What we mean is that such synthesis tools need to be complemented by tools of the 21st century, tools that allow for: (i) rapidly encompassing larger volumes of information without necessarily reading the complete book or paper; (ii) full transparency by providing the underlying dataset, thus building trust and creating no room for potential manipulation; and (iii) a possibility to correct and extend this dataset together with an automatic update of the synthesis results; to just mention three particularly important features.

Regarding the first feature, information graphics and visual analytics go in a promising direction (e.g. Keim *et al.*, 2010). The second feature is increasingly fulfilled, particularly by papers in journals that ask to provide the underlying data in freely accessible repositories. The third feature is challenging. With this book and the accompanying website at www.hi-knowledge.org, we are preparing for a first step along this path. The website is planned as an evolving platform to be improved and updated regularly. For example, the dataset will be corrected as soon as we become aware of mistakes (please point out errors you detect, e.g. by sending an e-mail). In the future, we aim to implement a Wiki system where registered users can correct data themselves. The dataset should also be extended in the future, so that more relevant data on each hypothesis can be included, e.g. publications that have become available after our systematic searches for each chapter were completed or publications that we missed in our searches. All changes in the data should automatically change the display of empirical support of the hypotheses.

Furthermore, we plan to extend the number of hypotheses covered. We have included in-depth analyses of a dozen invasion hypotheses in the previous chapters. An additional 23 invasion hypotheses were defined and plotted in a network of hypotheses in Chapter 7, this volume, which altogether covers 35 invasion hypotheses. There are even further hypotheses, e.g. the genome constraint hypothesis (Knight *et al.*, 2005), the AIAI hypothesis (anthropogenically induced adaptation to invade; Hufbauer *et al.*, 2012), the evolutionary imbalance hypothesis (Fridley and Sax, 2014), the invasive queens hypothesis (Platt and Jeschke, 2014), the environmental matching hypothesis (Iacarella *et al.*, 2015) or the intermediate distance hypothesis (Seebens *et al.*, 2017). Hence, there is room for extension.

Even more importantly, we suggest the network of hypotheses should grow beyond invasion biology and cover hypotheses from other disciplines and those hypotheses that bridge disciplines (as already suggested in Jeschke, 2014). This is possible with our framework because it is scalable to any set of hypotheses. A large hypothesis network, where each hypothesis is connected to empirical studies and underlying data, would be a possible realization of an atlas of scientific significance that Kitcher (2011) suggested in the above quote (see also Bollen *et al.*, 2009; Börner, 2010, 2015). The hypothesis network could thus develop into a powerful web portal with suggested settings for different user types such as scientists from different disciplines, students or decision makers and other stakeholders. We think of these settings as being customizable for the individual needs of each user who can change them and save these changes. For instance, a manager working on invasive fishes might be particularly interested in research focused on this group and thus filters the information on the website accordingly. She/he can log onto the website whenever she/he wishes and sees the updated evidence. In other words, we envision www.hi-knowledge.org as a web portal providing instantaneous, user-customized analyses based on a continuously updated database. As such, we hope it will become what could be called an atlas of invasion biology and later maybe even an atlas of science.

In the remainder of this chapter, we will link the hypothesis network from Chapter 7, this volume, with the empirical analyses in the previous chapters in order to: (i) colour code the network, thus hopefully allowing you, the reader, to rapidly grasp 'a lot of information without necessarily reading the complete book' (quoted from above); and (ii) present an improved hypothesis network in which the hypotheses are better empirically supported than in the original network. Such an improved network might help the field to abandon 'zombie' ideas more quickly (Fox, 2011; based on Quiggin, 2010) and to move forward more rapidly. We also address biases in the investigation of the 12 hypotheses covered here and whether empirical support for these hypotheses has changed over time.

A Hierarchically Structured Network of Invasion Hypotheses

Most of the 12 invasion hypotheses featured in the previous chapters were hierarchically divided into sub-(sub-)hypotheses, following the hierarchy-of-hypotheses (HoH) approach outlined in Chapters 2 and 6, this volume. Combining these hierarchically divided hypotheses with the hypothesis network provided in Chapter 7 leads to a hierarchically structured network of hypotheses. The first layer of this network includes the 12 major hypotheses, with the sub-(sub-)hypotheses being positioned in the secondary and tertiary layers. The HoHs in the previous chapters were already colour coded according to the level of empirical support; thus colour coding the first layer of the network is straightforward and, we believe, very powerful for quickly conveying the information concerning which hypotheses are well supported empirically and which are not (Fig. 17.1). In addition, Table 17.1 provides detailed information about the number of studies assessed for each hypothesis, and the percentage of studies being supportive, undecided and questioning. Overall, about 1100 studies were analysed in the previous chapters and form the empirical basis for Fig. 17.1.

For six of the 12 hypotheses, more than 50% of empirical studies were supportive.

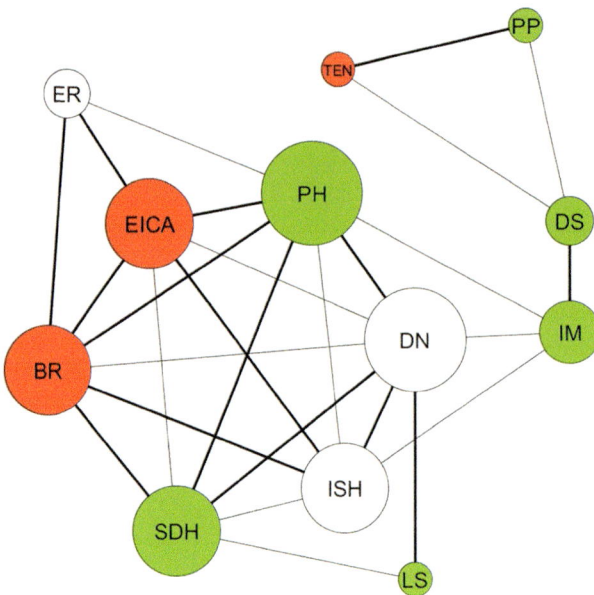

Fig. 17.1. The network of 12 invasion hypotheses featured in this book from Chapter 7 but colour coded according to the level of evidence reported for each major hypothesis, either based on meta-analysis (tens rule, Chapter 13) or the percentage of supporting or questioning studies (Chapters 8–12 and 14–16, cf. Table 17.1): green indicates that >50% of studies supported this hypothesis, red indicates that >50% of studies questioned this hypothesis and white indicates that there was no absolute majority of studies either supporting or questioning this hypothesis. Hypothesis names are abbreviated as follows: BR=biotic resistance, DN=Darwin's naturalization, DS=disturbance, EICA=evolution of increased competitive ability, ER=enemy release, IM=invasional meltdown, ISH=island susceptibility hypothesis, LS=limiting similarity, PH=plasticity hypothesis, PP=propagule pressure, SDH=shifting defence hypothesis and TEN=tens rule.

Table 17.1. Overview of the 12 invasion hypotheses featured in Chapters 8–16, together with the assessed numbers of studies (or, as indicated, numbers of investigated traits for Chapter 12), and percentage of these tests being supportive, undecided and questioning.

Hypothesis	No. studies	Supporting	Undecided	Questioning
Biotic resistance (BR)	155	30.3%	17.4%	52.3%
Darwin's naturalization (DN)	40	32.5%	17.5%	50.0%
Disturbance (DS)	126	58.7%	20.6%	20.6%
EICA	68 (no. traits)	41.2%	0.0%	58.8%
Enemy release (ER)	163	40.5%	33.1%	26.4%
Invasional meltdown (IM)	208	63.5%	11.5%	25.0%
Island susceptibility (ISH)	17	29.4%	23.5%	47.1%
Limiting similarity (LS)	33	57.6%	12.1%	30.3%
Plasticity hypothesis (PH)	115	60.9%	23.5%	15.7%
Propagule pressure (PP)	92	80.4%	7.6%	12.0%
Shifting defence (SDH)	68 (no. traits)	55.9%	0.0%	44.1%
Tens rule (TEN)	102	Different approach, see Chapter 13		
	$\Sigma=1118$ (duplicates that we are aware of were removed, i.e. for EICA and SDH, and DN and LS)			

The best-supported hypothesis was propagule pressure – here, about 80% of the studies were supportive. Numbers two and three in the list of best-supported hypotheses were the invasional meltdown and plasticity hypotheses. Three additional hypotheses were supported by the majority of available studies: disturbance, limiting similarity and shifting defence. Empirical data were ambiguous for three hypotheses: enemy release, Darwin's naturalization and island susceptibility. And empirical evidence clearly questioned the three remaining hypotheses: EICA, biotic resistance and the tens rule.

Interestingly, the EICA and biotic resistance hypotheses are showing up as neighbours in the hypothesis network (Fig. 17.1). They are thus somewhat similar. In particular, they do not explicitly and specifically consider the important role of humans and their actions with respect to biological invasions, at least compared to other invasion hypotheses such as the propagule pressure or disturbance hypotheses, which clearly highlight the roles of humans and are much better empirically supported. Furthermore, the biotic resistance hypothesis only looks at a property of the resident system but ignores properties of the introduced species (Chapter 8, this volume), and the tens rule does not offer any explanation or mechanistic insight (Chapter 13, this volume).

Revising or Abandoning Questioned Hypotheses: An Alternative Hypothesis Network

Given that three of the 12 invasion hypotheses are questioned by the majority of empirical evidence, it seems reasonable to revise or abandon them in order to allow the field to move forward. Holding on to poorly supported ideas clearly slows down or hinders progress (see also Jeschke *et al.*, 2012a). We will now take a closer look at the three empirically questioned hypotheses in order to decide whether there is hope in revising them or if they should better be abandoned.

In Chapter 12, this volume, the shifting defence hypothesis has been introduced as a revision of the EICA hypothesis. Although there still is potential to improve it from a conceptual point of view (see Chapter 12), the shifting defence hypothesis in its current version is backed by a comparably high percentage of studies. We therefore suggest abandoning the EICA hypothesis and using the shifting defence hypothesis instead.

Although the biotic resistance hypothesis is only poorly empirically supported, one of its sub-hypotheses seems to be a promising candidate to revise this classic idea. According to Chapter 8, this volume, those studies addressing the resistance hypothesis that looked at traits related to impact, such as abundance, biomass or cover of non-native species, showed higher levels of support than other studies of this hypothesis. On the basis of this finding, it was suggested to revise the resistance hypothesis as follows: 'ecosystems with high biodiversity are more resistant against non-native species than ecosystems with lower biodiversity, leading to lower levels of impact in highly diverse systems' (quoted from Chapter 8). We follow this suggestion and the proposed name as impact resistance hypothesis (IR).

Finally, Chapter 13, this volume, and references therein revealed that the tens rule is hardly empirically supported. Here as well, a revision was suggested, namely that the tens rule 'is replaced by two taxon-dependent hypotheses: the *50% invasion rule* for vertebrates and the *25% invasion rule* for other organisms, particularly plants and invertebrates' (quoted from Chapter 13).

We implemented these suggestions in Fig. 17.2, which shows a revised network of invasion hypotheses, now featuring 11 hypotheses. Seven of these hypotheses are empirically supported by the majority of available empirical studies, four hypotheses have received ambiguous empirical support and no hypothesis is questioned by available evidence (at least when taking the 50% threshold as done here, which can of course be discussed). Such an empirically supported conceptual backbone should be more useful for the further advancement of the discipline than the more weakly supported network illustrated in Fig. 17.1.

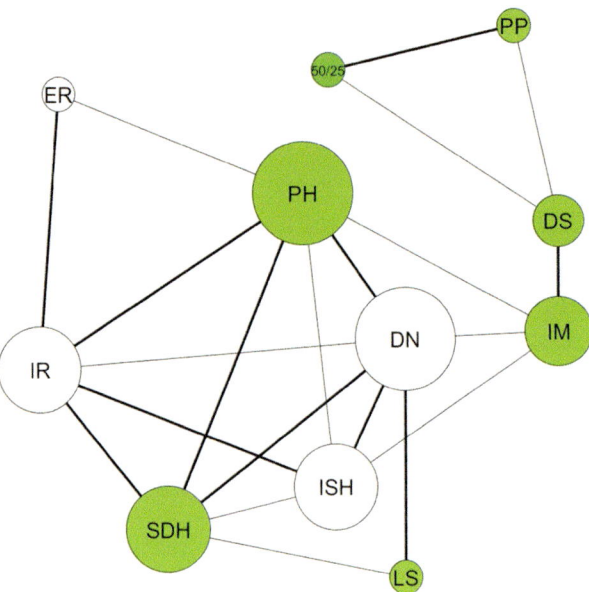

Fig. 17.2. Revised network of invasion hypotheses. Those hypotheses that are in red in Fig. 17.1 were either revised or abandoned based on the results in the corresponding chapters: the biotic resistance hypothesis (Chapter 8) was revised to the impact resistance hypothesis (IR), the tens rule (Chapter 13) was revised to the 50%/25% invasion rule (50/25), and the EICA hypothesis was abandoned because the shifting defence hypothesis (both featured in Chapter 12) is an advanced version of this hypothesis that is better empirically supported. The colour coding of the 11 hypotheses shown in this network is as for Fig. 17.1.

Research Biases: Most Studies have Focused on Terrestrial Plants Introduced to Affluent Countries

Any empirical evaluation of hypotheses based on a synthesis of the available data is only as good as the data. Biases in the data should thus be critically examined. Data for most hypotheses currently focus on terrestrial plants: plants are the focus of seven of the ten hypotheses for which we at least tried to collect data across taxonomic groups (Table 17.2). The chapter on the two remaining of the 12 hypotheses – EICA and shifting defence – did not even intend to include other taxonomic groups because these hypotheses have a strong focus on plants. Thus, only three of the dozen hypotheses featured in depth in this book, i.e. a quarter, have no focus on plants and three-quarters do. If we calculate the average across hypotheses, thereby weighting each hypothesis equally, 60% of the studies focus on plants,

followed by vertebrates with 24% and invertebrates with 14%, whereas all other taxa are basically ignored by invasion studies (see also Pyšek *et al.*, 2008; Jeschke *et al.*, 2012b) despite their high importance when it comes to invader impacts. It is well known that invasive pathogens such as the chytrid *Batrachochytrium dendrobatidis*, which threatens amphibians worldwide, as well as many plant pathogens have devastating effects (Bellard *et al.*, 2016; Lovett *et al.*, 2016). Although animals and other organisms usually have higher ecological and socio-economic impacts than plants (see also Vilà *et al.*, 2010), the majority of invasion studies focuses on plants.

Regarding the habitat, more than four out of five studies across all investigated hypotheses focus on terrestrial habitats (Table 17.3). Aquatic habitats have been largely ignored by invasion biologists, even though aquatic communities have frequently seen dramatic changes owing to

Table 17.2. Taxonomic coverage of the studies analysed in Chapters 8–11 and 13–16 on ten invasion hypotheses. The most frequently studied taxon is highlighted for each hypothesis. The EICA and shifting defence hypotheses (Chapter 12) are not included here because these focus on plants.

Hypothesis	Plants	Algae	Invertebrates	Vertebrates	Other
Biotic resistance (BR)	70.6%	2.5%	12.5%	13.1%	1.3%
Darwin's naturalization (DN)	69.2%	0.0%	2.6%	28.2%	0.0%
Disturbance (DS)	72.9%	3.1%	14.7%	9.3%	0.0%
Enemy release (ER)	74.8%	1.8%	11.7%	11.7%	0.0%
Invasional meltdown (IM)	36.4%	2.0%	38.4%	21.3%	2.0%
Island susceptibility (ISH)	35.3%	0.0%	0.0%	64.7%	0.0%
Limiting similarity (LS)	93.9%	0.0%	3.0%	3.0%	0.0%
Plasticity hypothesis (PH)	69.6%	0.0%	17.4%	13.0%	0.0%
Propagule pressure (PP)	44.6%	1.1%	21.7%	32.6%	0.0%
Tens rule (TEN)	32.3%	5.5%	18.1%	40.2%	3.9%
Average (each hypothesis weighted equally)	60.0%	1.6%	14.0%	23.7%	0.7%

Table 17.3. Major types of habitat covered by the studies analysed in Chapters 8–11 and 13–16 on ten invasion hypotheses. The most frequently studied habitat type is highlighted for each hypothesis. Chapter 12 on the EICA and shifting defence hypotheses is not included here because the whole chapter focused on terrestrial plants.

Hypothesis	Terrestrial	Freshwater	Marine
Biotic resistance (BR)	83.7%	9.2%	7.2%
Darwin's naturalization (DN)	90.0%	6.7%	3.3%
Disturbance (DS)	84.9%	4.0%	11.1%
Enemy release (ER)	82.8%	8.6%	8.6%
Invasional meltdown (IM)	63.5%	22.6%	13.9%
Island susceptibility (ISH)	82.4%	17.6%	0.0%
Limiting similarity (LS)	100.0%	0.0%	0.0%
Plasticity hypothesis (PH)	76.5%	19.1%	4.3%
Propagule pressure (PP)	77.6%	17.6%	4.7%
Tens rule (TEN)	74.2%	14.0%	11.8%
Average (each hypothesis weighted equally)	81.6%	11.9%	6.5%

invaders such as crayfish, crayfish plague, gobies, zebra or quagga mussels (e.g. Strayer, 2010; Gallardo *et al.*, 2016). Let us take the list of invasive alien species of European Union concern, which is connected to the EU Regulation No 1143/2014 and includes selected high-impact invaders. Of the 49 species currently listed there, about half are terrestrial and half are aquatic (depending on how one counts semi-aquatic–terrestrial species), although terrestrial invaders are much better investigated than aquatic invaders. It is reasonable to assume that the percentage of aquatic invaders on the list would be even higher if invasion biologists considered aquatic habitats more frequently. It is clear, however, that aquatic habitats finally need to get on the radar of many more invasion biologists than is currently the case. Also, the previous chapters demonstrated that most research in invasion biology is focused on affluent Western countries and continents (cf. Bellard and Jeschke, 2016); thus biases in the analysed data were observed in this respect as well.

Given that there are such biases in addressing invasion hypotheses, everyone who is interested in a particular taxon, habitat or geographic region should check whether the results for the selected taxon, habitat or region differ from the general pattern. For hypotheses with an overall high level of support, evidence for that specific study system could be weak, and vice versa, hypotheses with an overall low level of empirical support can be supported for certain taxa, habitats or regions (see e.g. the biotic resistance hypothesis in freshwater habitats, Chapter 8, this volume). An online tool such as the one planned to become available at www.hi-knowledge.org will allow for quick taxon-, habitat- and region-specific analyses. Until the online tool is available with this feature, everyone can download and check the datasets for their interests. Also, the previous chapters in this book have outlined where differences among taxonomic groups and habitats were observed for certain hypotheses. The database consisting of the approximately 1100 studies analysed here can be used in a multitude of ways, such as to quickly find examples for distinct mechanisms underlying biological invasions, because these are represented by different hypotheses. Hypotheses that are not supported in general might be highly relevant for particular study systems, e.g. for a given non-native species in a given region where it was introduced.

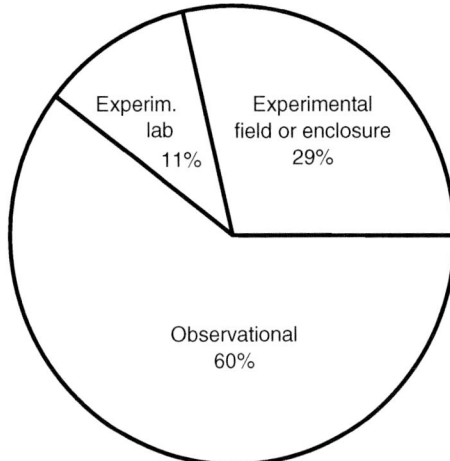

Fig. 17.3. Pie chart showing the percentage of studies applying different research methods, based on the combined data collected for Chapters 8–11 and 13–16 (n = 989 studies, all weighted equally; the sample size is smaller than in Table 17.1 because such data were not collected for Chapter 12 and studies with a mixed design were excluded).

Research Methods: Invasion Biology is Dominated by Observational Studies

The absolute majority (60%) of the studies analysed in the previous chapters were observational field studies (Fig. 17.3). Six out of ten studies were observational and only four out of ten studies were experimental, thus the discipline and most of its findings are based on observations with the known challenge that cause–effect relationships cannot really be inferred. Manipulative experiments, which probably dominate most other scientific disciplines, are in the minority in invasion biology. Field studies are vital for our understanding and are often observational by their very nature. Experiments, on the other side, allow for inferring causes and effects, yet may miss important factors and mechanisms acting in the field. If the results of observational studies differ from those of experimental studies, we suggest combining both approaches and thus working towards an understanding of the underlying reasons for these differences. In the case of the biotic resistance hypothesis (Chapter 8, this volume), observational studies provided significantly more questioning results than experimental studies at small spatial scales, possibly indicating that this hypothesis only holds for such scales – but studies combining observational with experimental tests in the same system should be conducted to clarify whether this is really the case. For the enemy release hypothesis (Chapter 11), we found the opposite: supporting evidence was mainly provided by observational studies, whereas experimental studies provided only rarely supporting evidence (a similar pattern was found for the disturbance hypothesis, Chapter 9, and the phenotypic plasticity hypothesis, Chapter 14). This might indicate that the current experiments do not capture all of the relevant mechanisms; again, studies combining field observation with experiments could be useful.

In general, the high percentage of observational studies in invasion biology might indicate that the discipline is still in a phase of exploratory research, scanning more for patterns than looking for mechanisms. The observational studies in invasion biology compiled in the previous chapters and available at www.hi-knowledge.org already present a strong empirical basis indicating some patterns, and we suggest they should now be used to perform well-designed experiments. We believe that a higher proportion of experimental studies in the field or in enclosures would help invasion biology to move forward efficiently.

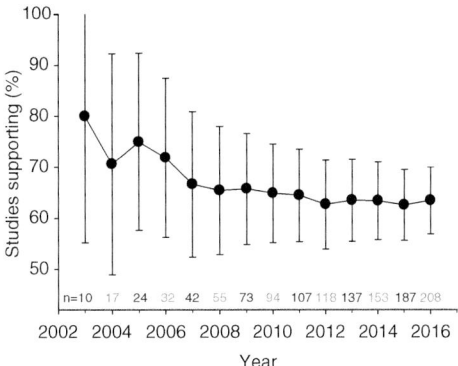

Fig. 17.4. Cumulative percentage of studies supporting the invasional meltdown hypothesis ±95% confidence intervals, using the dataset collected for Chapter 10 on this hypothesis; cumulative numbers of studies are indicated above the x-axis.

A Decline Effect for some, but not all Investigated Hypotheses

A temporal decline in the level of support for six out of six invasion hypotheses was reported by Jeschke *et al.* (2012a), and decline effects have been previously reported from a few other disciplines, particularly medicine, psychology and evolutionary ecology (Lehrer, 2010; Schooler, 2011). Such effects are critical because any evidence-based decision, e.g. by policy makers or managers, is highly time-sensitive. For instance, if a decision was taken on the basis of the tens rule some years ago to not increase border controls looking for species introductions, this decision should be overturned from today's perspective. The reasons behind decline effects include publication biases, biases in study organisms or systems and the psychology of researchers (Jeschke *et al.*, 2012a, and references therein).

Several of the previous chapters have found a similar temporal decline (Chapter 11 on enemy release, 13 on the tens rule, 14 on the plasticity hypothesis and 16 on the propagule pressure hypothesis); however, others did not (Chapters 8, 9 and 15). Chapters 10 and 12 have not addressed a potential decline effect. The dataset collected for Chapter 12 does not allow for the respective analysis but here we use the data collected

for Chapter 10 to check whether a decline effect was observed for the invasional meltdown hypothesis. We used the more than 200 studies included in this dataset to plot how the percentage of studies supporting the invasional meltdown hypothesis has changed over time (Fig. 17.4). Such a detailed time plot is particularly informative for large datasets with a number of studies for each year. The plot shows that the percentage of studies supporting the invasional meltdown hypothesis has declined over time, thus confirming the earlier finding by Jeschke *et al.* (2012a) who also found a decline effect for this hypothesis with a smaller dataset.

It is probably not surprising that not all hypotheses show a decline effect. For those that do show such an effect and where a sufficient number of studies have been published, a time plot such as Fig. 17.4 is informative because it might help assess whether the decline in support will continue further. In the case of the invasional meltdown hypothesis, 70–80% of the early studies were supportive. The frequency of supportive studies has decreased to a bit more than 60% over time but seems to be stable now; a further, strong decline therefore seems unlikely, at least if there is no abrupt change in how this hypothesis is addressed (cf. Chapter 10, this volume).

References

Bellard, C. and Jeschke, J.M. (2016) A spatial mismatch between invader impacts and research publications. *Conservation Biology* 30, 230–232.

Bellard, C., Genovesi, P. and Jeschke, J.M. (2016) Global patterns in threats to vertebrates by biological invasions. *Proceedings of the Royal Society B* 283, 20152454.

Bollen, J., Van de Sompel, H., Hagberg, A., Bettencourt, L., Chute, R., Rodriguez, M.A. and Balakireva, L. (2009) Clickstream data yields high-resolution maps of science. *PLoS ONE* 4, e4803.

Börner, K. (2010) *Atlas of Science: Visualizing what we Know*. MIT Press, Cambridge, Massachusetts.

Börner, K. (2015) *Atlas of Knowledge: Anyone can Map*. MIT Press, Cambridge, Massachusetts.

Fox, J. (2011) Zombie ideas in ecology. *Oikos blog*. Available at: oikosjournal.wordpress.com/ 2011/06/17/zombie-ideas-in-ecology (accessed 11 October 2017)

Fridley, J.D. and Sax, D.F. (2014) The imbalance of nature: revisiting a Darwinian framework for invasion biology. *Global Ecology and Biogeography* 23, 1157–1166.

Gallardo, B., Clavero, M., Sánchez, M.I. and Vilà, M. (2016) Global ecological impacts of invasive species in aquatic ecosystems. *Global Change Biology* 22, 151–163.

Hufbauer, R.A., Facon, B., Ravigné, V., Turgeon, J., Foucaud, J., Lee, C.E., Rey, O. and Estoup, A. (2012) Anthropogenically induced adaptation to invade (AIAI): contemporary adaptation to human-altered habitats within the native range can promote invasions. *Evolutionary Applications* 5, 89–101.

Iacarella, J.C., Dick, J.T.A., Alexander, M.E. and Ricciardi, A. (2015) Ecological impacts of invasive alien species along temperature gradients: testing the role of environmental matching. *Ecological Applications* 25, 706–716.

Jeschke, J.M. (2014) General hypotheses in invasion ecology. *Diversity and Distributions* 20, 1229–1234.

Jeschke, J.M., Gómez Aparicio, L., Haider, S., Heger, T., Lortie, C.J., Pyšek, P. and Strayer, D.L. (2012a) Support for major hypotheses in invasion biology is uneven and declining. *NeoBiota* 14, 1–20.

Jeschke, J.M., Gómez Aparicio, L., Haider, S., Heger, T., Lortie, C.J., Pyšek, P. and Strayer, D.L. (2012b) Taxonomic bias and lack of cross-taxonomic studies in invasion biology. *Frontiers in Ecology and the Environment* 10, 349–350.

Keim, D., Kohlhammer, J., Ellis, G. and Mansmann, F. (2010) *Mastering the Information age: Solving Problems with Visual Analytics*. Eurographics, Goslar.

Kitcher, P. (2011) *Science in a Democratic Society*. Prometheus, Amherst, New York.

Knight, C.A., Molinari, N.A. and Petrov, D.A. (2005) The large genome constraint hypothesis: evolution, ecology and phenotype. *Annals of Botany*, 95, 177–190.

Lehrer, J. (2010) The truth wears off. *New Yorker* Dec 13, 52–57.

Lokatis, S. and Jeschke, J.M. (2018) The island rule: An assessment of biases and research trends. *Journal of Biogeography* 45, 289–303.

Lovett, G., Weiss, M., Liebhold, A.M., Holmes, T.P., Leung, B., Fallon Lambert, K., Orwig, D.A., Campbell, F.T., Rosenthal, J., McCullough, D.G. *et al.* (2016) Nonnative forest insects and pathogens in the United States: Impacts and policy options. *Ecological Applications* 26, 1437–1455.

Platt, V. and Jeschke, J.M. (2014) Are exotic species red queens? *Ethology Ecology & Evolution* 26, 101–111.

Pyšek, P., Richardson, D.M., Pergl, J., Jarošík, V., Sixtová, Z. and Weber, E. (2008) Geographical and taxonomic biases in invasion ecology. *Trends in Ecology & Evolution* 23, 237–244.

Quiggin, J. (2010) *Zombie Economics: How Dead Ideas Still Walk Among Us*. Princeton University Press, Princeton, New Jersey.

Schooler, J. (2011) Unpublished results hide the decline effect. *Nature* 470, 437.

Seebens, H., Essl, F. and Blasius, B. (2017) The intermediate distance hypothesis of biological invasions. *Ecology Letters* 20, 158–165.

Strayer, D.L. (2010) Alien species in fresh waters: ecological effects, interactions with other stressors, and prospects for the future. *Freshwater Biology* 55 (Suppl. 1), 152–174.

Vilà, M., Basnou, C., Pyšek, P., Josefsson, M., Genovesi, P., Gollasch, S., Nentwig, W., Olenin, S., Roques, A., Roy, D., Hulme, P.E. and DAISIE partners (2010) How well do we understand the impacts of alien species on ecosystem services? A pan-European, cross-taxa assessment. *Frontiers in Ecology and the Environment* 8, 135–144.

18

Conclusions and Outlook

Tina Heger[1,2,3]* and Jonathan M. Jeschke[4,5,3]

[1]University of Potsdam, Biodiversity Research/Systematic Botany, Potsdam, Germany; [2]Technical University of Munich, Restoration Ecology, Freising, Germany; [3]Berlin-Brandenburg Institute of Advanced Biodiversity Research (BBIB), Berlin, Germany; [4]Leibniz-Institute of Freshwater Ecology and Inland Fisheries (IGB), Berlin, Germany; [5]Freie Universität Berlin, Institute of Biology, Berlin, Germany

Abstract

In Chapters 8–16, the hierarchy-of-hypotheses (HoH) toolbox has been applied in various ways. The results displayed as HoHs and additional bar graphs in most chapters, and as networks in Chapter 17, can now be used to identify knowledge gaps and promising paths for future research. We suggest some further steps that could be taken, making use of the data gathered for this book (which are freely available at www. hi-knowledge.org) and possible avenues for more general methodological advancements. To strengthen research in invasion biology, we specifically suggest four steps: (i) a more efficient provision and use of existing data and knowledge; (ii) a stronger focus on experiments or ideally research projects combining field studies with experiments; (iii) the development of sets of conceptually based 'case-sensitive' generalizations; and (iv) continuous integration of newly gained data and knowledge into a growing and developing *atlas of invasion biology*, which could potentially be expanded into an *atlas of science*.

In this final chapter, we will first discuss how the chapters in Part II of the book made use of the HoH approach, what we think was achieved and which further methodological steps can be taken. The second section of this chapter, and final section of the book, will be an overall brief conclusion about the current state of invasion biology and possible ways forward.

Concluding Considerations on the Hierarchy-of-hypotheses Approach

The HoH toolbox: tools and applications

In Chapter 2, we outlined the hierarchy-of-hypotheses (HoH) approach and formulated as one aim of this book to test its usefulness as a conceptual backbone for invasion biology. In Chapter 6, the HoH approach was described as a toolbox: the core tool of using a hierarchical structure can be applied to different objects (e.g. hypotheses, predictions and concepts), and can be supplemented by a systematic review, meta-analysis *sensu lato* using either a semi- or fully quantitative approach, by a weighting procedure or other approaches. In Chapters 8 to 16, these tools were used and combined in different ways (Table 18.1). In all of these chapters, a systematic review of the available literature was performed. Chapters 8–10 and 14–16 then used the semi-quantitative approach of

* Corresponding author. E-mail: tina-heger@web.de

© CAB International 2018. *Invasion Biology: Hypotheses and Evidence*
(eds J.M. Jeschke and T. Heger)   167

Table 18.1. Overview of which of the tools proposed in Chapters 2 and 6 have been used in each chapter.

Tool	Ch. 7 Network	Ch. 8 BR, ISH	Ch. 9, DS	Ch. 10, IM	Ch. 11, ER	Ch. 12, EICA, SDH	Ch. 13, TEN	Ch. 14, PH	Ch. 15, DN, LS	Ch. 16, PP
Systematic review		×	×	×	×	×	×	×	×	×
Semi-quantitative assessment (levels: supporting, questioning, undecided)		×	×	×	×		×	×	×	×
Fully quantitative meta-analysis							×			
Weighting of studies based on methodological design				×						
Closer assessment of how methodological design affects results (Levins' trade off)					×					
Comparison of taxa and habitats		×	×	×	×		×	×	×	×
HoH colour coded with 50% evidence threshold		×	×	×				×	×	×
HoH with level of evidence as bar graphs					×					
Network of hypotheses	×									

BR = biotic resistance, DN = Darwin's naturalization, DS = disturbance, EICA = evolution of increased competitive ability, ER = enemy release, IM = invasional meltdown, ISH = island susceptibility hypothesis, LS = limiting similarity, PH = plasticity hypothesis, PP = propagule pressure, SDH = shifting defence hypothesis and TEN = tens rule.

classifying evidence provided by the analysed empirical studies into supporting, questioning or undecided studies. The results were displayed as a graphical representation of the underlying HoH, and most authors used a colour code to illustrate the overall level of evidence for every hypothesis at each hierarchical level. Colours were assigned according to the percentage of studies supporting (green) or questioning (red) the respective hypothesis, using a 50% threshold. White boxes indicated inconclusive results or sample sizes lower than five. In Chapter 11, the semi-quantitative approach was applied in the same way but, for the graphical display, overall evidence was illustrated as bar graphs giving 'raw' percentages, without applying the 50% threshold. Chapter 13 applied a fully quantitative meta-analysis *sensu stricto* instead of a semi-quantitative approach. When displaying the results, the colour code in the HoH has been chosen to mirror the results of this fully quantitative analysis. In Chapters 8–11 and 14–16, the different methodological

approaches of the analysed empirical tests have been taken into account for additional analyses. Furthermore in Chapter 10, studies were weighted on the basis of their methodological approach and, in Chapter 11, differences in methodological approaches were analysed in more detail. In each chapter, the authors discussed if the focal hypothesis should be revised given the results they found.

In Chapter 7, a network of hypotheses has been introduced, showing which major hypotheses in invasion biology are similar based on the percentages of characteristics (or 'traits') that they share. Alternative approaches for creating hypothesis networks were mentioned there as well. This chapter focused on the top level of the hierarchy and showed connections for the major hypotheses featured in this book. In Chapter 17, we combined this network with results from Chapters 8–16, and offered a revised network in which the three hypotheses that had received low empirical support were replaced or removed.

The HoH toolbox therefore has been used in diverse ways in this book. We believe that the results can be very helpful: the graphical displays of the detailed HoHs provide a quick overview on which aspects of the respective hypothesis have been tested and what was the level of empirical evidence the tests produced. In combination with the bar graphs shown in most chapters, they also help uncover at least parts of the underlying complexity. This information, potentially even more helpful as an online resource at www.hi-knowlege.org, can be, for example, used to decide on future basic or applied research projects. We further hope that the network showing how the major hypotheses are connected with each other stimulates conceptual and empirical research exploring these connections, for instance empirical studies that address and compare two or more connected hypotheses.

Where to go from here?
Potential next steps

There are many more possibilities to use the HoH approach that have not been fully explored here. For example, in addition to experimental and observational data, results from modelling studies should also be used in the future to assess the level of support for different (sub)hypotheses. Furthermore, the analyses presented in this book could be used to further refine the presented hypotheses. So far, we mainly suggested refinements based on content (e.g. to use the shifting defence hypothesis rather than the less supported evolution of increased competitive ability (EICA) hypothesis); but based on the information provided here and at hi-knowledge.org, it would also be possible to refine the hypotheses further, for example regarding habitats. For the biotic resistance hypothesis, we found that it is more frequently supported in aquatic than terrestrial habitats; hence rather than revising it according to content (as for the impact resistance hypothesis), one could also revise it by discriminating major habitat types. Significant differences in the level of empirical

support between taxonomic groups were found for both the invasional meltdown hypothesis and the tens rule. For the latter, the suggested revision took these differences into account. Revising hypotheses not only makes sense for those that are poorly empirically supported but also for those with differing levels of evidence across subhypotheses. Such revisions would allow for a better and systematic understanding of the domain in which each hypothesis is applicable. They would lead to a set of empirically driven generalizations with clearly defined boundaries (as in Murray, 2004; see Chapter 6, this volume), making it possible to derive predictions 'detached from generality' in the sense that they take into account context dependence (Evans *et al.*, 2013).

An alternative or supplementary aim could be to search for a theoretically–conceptually driven set of generalizations. The scope of this book was to confront existing hypotheses with evidence reported in empirical studies. In relation to Fig. 6.1 in Chapter 6, the applications of the HoH approach featured in this book are located at the evidence-driven side of the gradient we showed there: they are designed to structure evidence. One can alternatively use the HoH approach to structure research or even theory. Instead of using existing data to look for research systems or classes of cases in which a major hypothesis has been addressed, the same major hypothesis can be hierarchically divided into sub-hypotheses based on conceptual considerations. For the enemy release hypotheses, for instance, one can ask which groups of species are especially likely to profit from enemy release (see Chapter 2). Here the aim would be to find classes of cases in which the same basic mechanisms apply and thus to develop a set of conceptually based 'case-sensitive' generalizations.

These ideas for future research should, however, not obscure what has been reached already. We hope readers agree that the HoH approach did indeed prove useful for invasion biology. Therefore, as we indicated in Chapter 2, we hope to see it also being applied in other disciplines. In some occasions, this has already happened: Gibson (2015) used it in a textbook for plant population biology, Dietl (2015) applied it within palaeontology, Lokatis and Jeschke (2018) applied it within biogeography, and there are preliminary results of an application in cancer research (Bartram and Jeschke, unpublished).

A potential limitation of applying the HoH approach in the same way as in this book to other disciplines is that not all disciplines are as strongly structured according to major hypotheses as is the case for invasion biology. In such cases, structuring theory or empirical evidence hierarchically seems to be less straightforward, but may still be useful, e.g. according to the major research questions driving a discipline. Also, in any discipline the approach might, for example, be used to structure methods or predictions.

Ideas for research synthesis in the 21st century

The question we posed in Chapter 2 of this book was: 'which major hypotheses in invasion biology are backed up by empirical evidence, and which should be re-formulated or completely discarded?' In Chapters 8–17, we made several suggestions on how to answer this question. Most of these suggestions are based on the notion that a hypothesis is not useful if it is questioned by the majority of existing studies, yet we pointed out in Chapters 2 and 6 that this is debatable.

A very important next step from our perspective is therefore to explore further the philosophical implications of the HoH approach, or of current habits of hypothesis testing in ecology and science in general. Experts on conceptual theory, philosophy of science and statistical analysis should, in a joint effort, develop a strong basis for research synthesis and theory development. Current synthesis in ecology has in fact already moved away from Popperian falsificationism but this development is currently happening without much reflection on its philosophical implications. Chapters 4, 6 and 12 in our opinion show that exchange between ecologists and philosophers can be fruitful. To explicitly and knowingly leave behind naïve falsificationism and find new ways to reach strong explanations (in the sense of Bartelborth, 2007) and predictions by considering context dependencies should be a major aim for the near future. Agreement on what a strong enough explanation is, which level of supporting evidence can be demanded, or how robust a theory should be (see Chapters 4, 6 and 12) would strongly aid the creation of an 'atlas of scientific significance' (see Chapter 17).

One way to construct such an atlas for invasion biology could be to design a visual representation of the hierarchical network of hypotheses as suggested in the previous chapter. We regard the HoH approach combined with network analyses as a first example of alternative ways to handle, compare and connect hypotheses. Implemented as an online tool at hi-knowledge.org, we expect it to facilitate the development of an atlas of scientific significance. As indicated in the previous chapter and elsewhere in this book, the construction of such an atlas is nevertheless challenging. For example, a full representation of a structured network would not only include connections at the highest level of hypotheses, but additionally on the lower levels, including connections between different major hypotheses, potentially connecting them at all hierarchical levels. We envision hi-knowledge.org as a platform that will provide such a structured network. We think of this web portal as a tool for continuous, web-aided synthesis, where evidence is constantly updated and fed into the growing network, displaying the level of evidence for every branch in each hierarchy (see Chapter 17).

To move further in this direction, it will be necessary to discuss very concrete and problem-oriented questions. How should data in the future be integrated in the database: by means of a Wiki system where the researchers who performed a study enter their data? Which incentives are needed to motivate researchers to contribute? Which statistical tools can help build and maintain the network? What should the graphical display look like? Clearly, such questions need to be tackled jointly with web designers, database managers, biostatisticians, network experts, philosophers, artists and others.

In a nutshell, we envision that future work outlined above leads towards a system of generalizations with known boundaries of their applicability, thus allowing context-specific predictions. These generalizations will be derived using a combination of systematic, conceptual work and synthesis of empirical and modelling studies, possibly using the HoH approach in combination with network analysis. The underlying theory and empirical data will be openly accessible, easy to comprehend owing to good visualization tools, and constantly updated as new data and information become available. Starting from invasion biology, this novel way of research synthesis will expand towards other disciplines.

Hypotheses and Evidence in Invasion Biology: Conclusions

Invasion biology has arguably 'come of age' (see Chapter 1), and the information gathered in the previous chapters conveys much good news. From a nature conservation point of view, it is, for example, good to know that there is comparably strong support for the disturbance hypothesis because this gives hope that measures aiming at the prevention of invasions by minimizing disturbance in sensible areas can be successful. Also, the overall positive evidence for the limiting similarity hypothesis can be used as an incentive to further develop measures to prevent alien species establishment during restoration (e.g. Yannelli *et al.*, 2017). That there is so much positive evidence for the propagule pressure hypothesis confirms current developments to focus on pathway management (e.g. European Union, 2014). High levels of supporting evidence for these and three other hypotheses (plasticity, invasional meltdown and shifting defence) indicate that we really are beginning to get a handle on the mechanisms that drive invasions. The HoH approach with its possibilities for visualization of the level of evidence

of different 'branches' of a hypothesis seems to be able to account for the complexity and context dependence of invasion processes.

Nevertheless, we think the previous chapters also suggest that invasion biology is still far from being a ripe discipline. Like many young adults, it is still in its 'exploratory phase', indicated by the prevalence of observational studies (see Chapter 17). Also, the overview presented in this previous chapter does not only carry good news; admittedly, in a way it is also sobering: for several hypotheses there is more negative than positive evidence and, even for the hypothesis with the highest level of empirical support (the propagule pressure hypothesis, Chapter 16), still 12% of the empirical studies delivered questioning results. These results call for leaving at least some of the beaten paths and looking for alternative ways to proceed.

In the previous paragraphs and chapters, we suggested several measures that can be implemented to advance invasion biology. Arguably, the four most important ones are:

1. A more efficient provision and use of existing data and knowledge; e.g. by integrating synthesis tools such as the Wiki-based hierarchical networks proposed in Chapter 17 into existing data repositories.
2. A stronger focus on experiments, or ideally research projects combining observations and experiments done in the field and controlled environments, addressing gaps in existing knowledge.
3. The development of sets of conceptually based 'case-sensitive' generalizations.
4. Continuous integration of all newly gained data and knowledge into a growing and freely available atlas of invasion biology.

We are living in exciting times where huge amounts of data are being gathered every moment. It is a big and important task to better synthesize these data and to further our actual knowledge and understanding of the world. We hope this book and the web portal hi-knowledge.org will stimulate discussions and research directed at this grand task, in the context of biological invasions and beyond.

References

Bartelborth, T. (2007) *Erklären*. De Gruyter, Berlin, Germany.

Dietl, G.P. (2015) Evaluating the strength of escalation as a research program. *Geological Society of America Abstracts with Programs* 47, 427.

European Union (2014) Verordnung (EU) Nr. 1143/2014 des Europäischen Parlaments und es Rates vm 22. Oktober 2014 über die Prävention und das Management der Einbringung und Ausbreitung invasiver Arten.

Evans, M.R., Grimm, V., Johst, K., Knuuttila, T., de Langhe, R., Lessells, C.M., Merz, M., O'Malley, M.A., Orzack, S.H., Weisberg, M., Wilkinson, D.J., Wolkenhauer, O. and Benton, T.G. (2013) Do simple models lead to generality in ecology? *Trends in Ecology & Evolution* 28, 578–583.

Gibson, D. J. (2015) *Methods in Comparative Plant Population Ecology*. Oxford University Press, Oxford, UK.

Murray, B.G.J. (2004) Laws, hypotheses, guesses. *The American Biologist Teacher* 66, 598–599.

Yannelli, F.A., Hughes, P. and Kollmann, J. (2017) Preventing plant invasions at early stages of revegetation: The role of limiting similarity in seed size and seed density. *Ecological Engineering* 100, 286–290.

Index

This book is linked to a companion website
www.hi-knowledge.org